Heavy Hydrocarbon Resources

ACS SYMPOSIUM SERIES **895**

Heavy Hydrocarbon Resources

Characterization, Upgrading, and Utilization

Masakatsu Nomura, Editor
Osaka University

Parviz M. Rahimi, Editor
National Centre for Upgrading Technology

Omer Refa Koseoglu, Editor
Saudi Aramco

Sponsored by the
ACS Divisions of Fuel Chemistry and
Petroleum Chemistry, Inc.

American Chemical Society, Washington, DC

Library of Congress Cataloging-in-Publication Data

Heavy hydrocarbon resources : characterization, upgrading, and utilization / Masakatsu Nomura, editor, Parviz M. Rahimi, editor, Omer Refa Koseoglu, editor.

    p. cm.—(ACS symposium series ; 895)

    "Developed from a symposium sponsored by the Divisions of Fuel Chemistry and Petroleum Chemistry, Inc. at the 225th National Meeting of the American Chemical Society, New Orleans, Louisiana, March 23–27, 2003..."--Pref.

    Includes bibliographical references and index.

    ISBN 0–8412–3879–0 (alk. paper)

    1. Asphaltene—Congresses. 2. Petroleum—Congresses.

    I. Nomura, Masakatsu, 1940- II. Rahimi, Parviz M., 1946- III. Koseoglu, Omer Refa, 1956- IV. American Chemical Society. Meeting (225th : 2003 : New Orleans, La.) V. Series.

QD305.H7H39   2004
547′.412—dc22

         2004051025

The paper used in this publication meets the minimum requirements of American National Standard for Information Sciences—Permanence of Paper for Printed Library Materials, ANSI Z39.48–1984.

PRINTED IN THE UNITED STATES OF AMERICA

Foreword

The ACS Symposium Series was first published in 1974 to provide a mechanism for publishing symposia quickly in book form. The purpose of the series is to publish timely, comprehensive books developed from ACS sponsored symposia based on current scientific research. Occasionally, books are developed from symposia sponsored by other organizations when the topic is of keen interest to the chemistry audience.

Before agreeing to publish a book, the proposed table of contents is reviewed for appropriate and comprehensive coverage and for interest to the audience. Some papers may be excluded to better focus the book; others may be added to provide comprehensiveness. When appropriate, overview or introductory chapters are added. Drafts of chapters are peer-reviewed prior to final acceptance or rejection, and manuscripts are prepared in camera-ready format.

As a rule, only original research papers and original review papers are included in the volumes. Verbatim reproductions of previously published papers are not accepted.

ACS Books Department

Contents

Indexes

Preface

Because o f a r ising g lobal c oncern f or t he e nvironment, m any governments have been instituting increasingly stringent fuel specifications. These specifications, however, are b ecoming d ifficult t o m eet where crude oil quality is decreasing worldwide, resulting in increasing volumes of lower quality residues. Characterization of petroleum-derived asphaltenes presents researchers with a great challenge that may help solve the complexity of heavy hydrocarbon materials; the governing of asphaltenes is believed to be the key in suppressing fouling in oil production and coking during the pyrolysis and hydroprocessing of oil. Understanding these complicated materials is beneficial both in solving the causes of fouling in production and coke formation in conversion processes.

In March of 2003, the American Chemical Society (ACS) annual meeting was held in New Orleans, Louisiana where a session titled *Heavy Hydrocarbon Resources: Characterization, Upgrading and Utilization* was chaired by the editors of this ACS symposium series book. More than 18 original papers were submitted and very fruitful discussions were held. In the early stages of planning this session when Parviz Rahimi and Refa Koseoglu kindly joined in the planning and implementation, I had some idea that we would have little problems in gathering e nough h igh-quality r esearch p apers b ecause, a t that time in Japan, on relatively large-scale joint study (that included researchers in the United States and Japan) on asphaltenes already existed. The details of this project issued on R&D related to conversion technologies based on a structural study of petroleum-derived asphaltenes subsidized by New Energy and Industrial Technology Development Organization (NEDO).

Rahimi, Koseoglu, and I were contacted by ACS concerning the possibility of publishing these papers in book form. I contacted almost all the presenters about this possibility, and their reactions were very positive. We are pleased to present these papers to you now.

After overcoming several barriers, we asked 14 authors to submit their papers that were then sent to reviewers. During this period, we editors conducted an exhaustive survey of asphaltene-related papers during the past five years and found more than 100. These papers were separated into three categories—for example, properties and aggregation, structure and aggregation, and utilization. Each editor reviewed one-third of these papers as a means of reviewing the most recent developments in asphaltene studies; a review of 57 papers among them is included in the first chapter of this book. If there are some ambiguities in this chapter, it is my responsibility. We recommend that readers examine this first chapter's summary prior to reading later chapters.

Chapters 2 to 15 handle such wide-ranging and fascinating topics as the chemical constituents of asphaltenes, aggregation characteristics, dissolution properties of asphaltenes and resins, density simulations of asphaltenes, carbonization reactivity, thermal properties, coking reactiveity, characterizations of catalysts when hydroprocessing asphaltenes, and a newly built catalyst system for hydrocracking of asphaltenes. Here, the readers will find many interesting results concerning asphaltene chem.-istry and asphaltene properties, as well as useful and indispensable scientific background for R&D into asphaltene processing and conversion technologies.

Finally, I express my sincere thanks to Miyako Inoue for her devoted efforts in collecting and handling many papers from respective researchers. My sincere thanks go to Satoru Murata for his thorough review of references cited in Chapter 1.

Masakatsu Nomura
Emeritus Professor
Department of Applied Chemistry
Faculty of Engineering
Osaka University
2–1 Yamada-oka, Suita
Osaka 565–0871, Japan

Heavy Hydrocarbon Resources

Chapter 1

Advances in Characterization and Utilization of Asphaltenes

Masakatsu Nomura[1], Parviz M. Rahimi[2], and Omer Refa Koseoglu[3]

[1]Department of Molecular Chemistry, Graduate School of Engineering,
Osaka University, Yamada-oka, Suita, Osaka 565–0871, Japan
[2]National Centre for Upgrading Technology (NCUT),
1 Oil Patch Drive, Suite A202, Devon, Alberta T9G 1A8, Canada
[3]Research and Development Center, Saudi Aramco,
P.O. Box 62, Dhahran 31311, Saudi Arabia

A survey of the research related to petroleum asphaltenes, published over the past five years, yielded about 100 papers. The three authors of this chapter each examined the papers related most directly to their areas of specialization. 57 papers were selected as references and their results have been summarized in three categories: asphaltene properties, asphaltene structure, and utilization. Properties and structure were discussed in view of a commonly accepted characterization of asphaltenes, that is, aggregation. Utilization was discussed in terms of catalyst development studies, pyrolysis and hydroprocessing, while reactive characterization was important to the relationship between properties and reactivities of asphaltene.

Introduction

"Asphaltenes" are complex hydrocarbon mixtures that constitute a significant portion of many petroleum feedstocks. They are believed to cause fouling during production operations and are a cause for coke formation in conversion processes. Therefore, asphaltenes have been the subject of many investigations over the years.

In this introductory chapter, the three editors of this book sum up recent progress from more than fifty papers on asphaltene chemistry, and the research and development of utilization technologies concerning asphaltene substances.

Asphaltene properties and aggregation

To begin with, we have reviewed, to the best of our ability, the latest literature on asphaltene properties. It is interesting to note that even after many years of study on asphaltene properties and their chemistry, researchers are still investigating which method for asphaltene separation gives consistent results. For instance Yarranton et al. (1) have shown that separation of asphaltenes and their properties is very sensitive to the method of preparation. It was shown that after separation and filtration using conventional IP or ASTM methods, the extent of washing the filter cake with solvent can yield asphaltenes with different properties. Repeated washing can remove more resinous materials; however, there is no consistency in properties among the asphaltenes produced. The Soxhlet extraction technique proved to be the method sufficient to produce asphaltenes with consistent properties.

Because of the complexity of the molecular composition of petroleum, the scientific community has accepted the solubility class. Asphaltenes, the least soluble component of the oil, exist in colloidal aggregate and are insoluble in alkanes. The soluble part is called maltenes, which are composed of saturates, aromatics and heavier component resins. Although, in some respects, the borderline between asphaltenes and resins is a matter of definition, solvent classification remains a useful tool to categorize the heavy components in petroleum according to their solubilites. Vast numbers of studies reported in the literature are devoted to the properties of asphaltenes and their solubility. However, there is often different interpretations of experimental results, partly because of the complexity of the molecular structure of asphaltenes. To postulate a detailed asphaltene precipitation mechanism based on a vague and unknown molecular structure can only lead to inaccurate conclusions for asphaltene precipitation.

The currently accepted colloidal petroleum model requires that resin molecules act as asphaltene stabilizers and prevent asphaltenes from precipitating. Steric stabilization by resins of asphaltene aggregates has been invoked by Mansoori et al. (2) However, recent experimental evidence has challenged the view that resins provide steric stabilization of asphaltene aggregates by attaching to their periphery and stopping flocculation. It has been shown that resin solutions in n-heptane can adsorb on the surface of asphaltenes leading to multilayer formations. (3) Further, in another study by Gutiérrez et al., (4) it was demonstrated that, whereas the whole asphaltenes from Cerro Negro were soluble in toluene, about 70 % was insoluble. In another words, 30 % of the asphaltenes playing the role of resin kept the other 70 % in the solution. Based on these findings, a simpler model for asphaltene colloids in toluene was proposed by León et al. (5) In this model, insoluble asphaltenes constitute the core and the soluble asphaltenes fraction that is exposed to solvent molecules is attached to the periphery of the insoluble asphaltenes. These results led the authors further to investigate the interaction of asphaltenes with the native resins of Venezuelan crudes. Adsorption isotherms of two resins on two different asphaltenes were studied using a UV-vis spectrophotometric technique. At the same time, the stabilizing power of resins was compared with the known amphiphiles of an asphaltenes stabilizer such as nonylphenol. It was shown that resins adsorb and form a multi-layer structure on the surface of asphaltenes. When resins were added to the asphaltene sample immersed in n-heptane, a volumetric expansion in asphaltenes was observed, indicating that resins penetrated into the microporous structure of the asphaltenes. By comparing the stabilizing power of amphiphiles with native resins, it was shown that at the same equilibrium concentration the native resins adsorb in a lower amount than amphiphiles. However, some native resins were capable in dissolving more asphaltenes and were more effective as asphaltenes stabilizers. Based on these results the authors proposed a model for asphaltenes stabilization by resins. According to this model: a) resins are firstly adsorb on the surface of asphaltenes; b) the resins penetrate into the microporous structure of the asphaltenes; c) the resins break the microporous of the asphaltenes; and d) the asphaltenes-resins particles diffuse in the solvent.

The complex composition of petroleum changes during production, transportation and processing. These changes in composition have a significant effect on the stability of asphaltenes. In some research, the widely accepted stabilization of asphaltenes with resins has been questioned. (6) It has been pointed out that for asphaltenes stabilization in oil there is no need for a special interaction between asphaltenes and resins. Furthermore, the flocculation of asphaltenes occurs because of non-polar van der Waals forces. (7) In a number of analytical methods for the determination of onset solubility parameters, the oil is diluted with toluene and then titrated with a hydrocarbon (heptane). It has been assumed that the presence of toluene has no effect on solubility parameters but this has never been verified. Stability of asphaltenes in solutions can be quantified at the onset of asphaltenes flocculation by measuring the solubility parameters of the mixtures. An alternative method using a refractive index has been proposed as an indicator of asphaltene stability. (7,8)

Using a refractive index technique, the onset of asphaltene precipitation for a number of crudes was determined. It was shown the refractive index at the asphaltene flocculation point was dependent on the crude type. (9) It was further shown that there is a relationship between the carbon number of the flocculating agents and their molar volume in $(mL/mole)^{1/2}$, $(v_p^{1/2})$. Using this relationship, it is possible to determine, for a given oil, the size of hydrocarbon that causes asphaltene precipitation. Naturally occurring n-alkanes can destabilize asphaltenes. It was found that the live oils containing naturally occurring n-alkanes greater than 28 had no aggregates, whereas the aggregates were present in oils containing alkanes less than 28.

Dilution of crude with toluene prior to titration with n-alkanes resulted in a decrease in the refractive index at the flocculation onset. This means that the solubility parameters determined from the dilute solutions of oil were not accurate since the solubility of asphaltenes at the flocculation point had changed by the addition of toluene.

It has also been argued that resins have no special effect on stabilizing asphaltenes. When resins from one oil (in the range of 1-5 %) were added to toluene solution of asphaltenes from a different oil, only incremental improvements in asphaltene stability were observed. These results may indicate that resins merely act as good solvating agents and there are no special interactions between asphaltenes and resins.

The onset of asphaltene precipitation has been shown to change with the carbon number of the flocculating agent. (10) As the carbon number of the n-alkane is increased from C5 to C7 the onset of asphaltene precipitation for Rangely crude increased. Moreover, the addition of a long chain alkane such as eicosane and tetracosane accelerated asphaltene precipitation whereas addition of naphthalene and phenanthrene delayed precipitation.

Other materials such as CO_2 can also initiate asphaltene precipitation. The effect of CO_2 pressure on asphaltene precipitation from a Chinese oil field was examined under the reservoir conditions. (11) It was shown that, without CO_2, injection pressure depletion resulted in no precipitation. However, in the presence of CO_2, asphaltene deposition was dependent on the CO_2 pressure. It was found that if the CO_2 pressure was below the minimum miscibility pressure (MMP) no asphaltenes were precipitated and pressure higher than MMP was required to initiate asphaltene deposition. In another study, Idem et al. (12) investigated the kinetic of CO_2-induced asphaltene precipitation of three Saskatchewan crude oils. It was shown that CO_2, as a precipitating solvent, was more effective in oils with high API and paraffinic content than in oils with high aromaticity and asphaltene content. The order of reaction with respect to CO_2 concentration and asphaltenes content was very high and was shown to be temperature dependent. The high reaction order implies that asphaltene precipitation by CO_2 may not be the primary process. Since the reaction order with respect to CO_2 was higher than the reaction order for asphaltenes, it was concluded that asphaltene precipitation was more sensitive to CO_2 addition rather than asphaltene content. Recently, Vafaie-Sefti et al. (13) researched further in predicting asphaltene precipitation in crude oil during production. Their model is able to predict the effect of pressure and solvent composition on asphaltene precipitation. In this study the variation in CO_2 pressure as well as the amount and addition of resin on asphaltene precipitation was investigated. It was shown that as the pressure increased at constant temperature the amount of asphaltene precipitation increased and then decreased at higher pressures. The increase in the amount of CO_2 at constant pressure and temperature produced more asphaltene deposits and reached a plateau. The addition of resins to crude had a positive effect and resulted in reducing asphaltene precipitation. If the addition of resins continued, at a certain concentration, asphaltene deposition was significantly increased. It is believed that the resins themselves start precipitating.

Asphaltene structure and aggregation

The presence of asphaltenes in petroleum can cause problems during production, transportation and processing in the refinery. Asphaltene precipitation can be initiated by the addition of paraffinic solvents. During thermal or catalytic processing of petroleum, where the protecting resin layer is destroyed, asphaltenes can precipitate. Asphaltene precipitation on the surface of a catalyst can cause serious deactivation during hydrotreating, reducing sulphur and nitrogen removal. Detailed structural analysis of asphaltenes would help for a better catalyst design. What is known is that asphaltenes contain some fused aromatic rings (core) bearing heteroatoms such as O, N, S and metals, including V, Ni and Fe. Aromatic rings may carry aliphatic chains with variable lengthes. Even the positions of functional groups and saturated chains relative to the core, and the degree of aromatic condensation, are speculative at best. More than 25 papers were examined to gain insight into recent ideas about asphaltene structure. Gray (14) published an interesting paper concerning consistency of asphaltenes' chemical structures and pyrolysis behavior. He depicted two types of asphaltene structures, including a condensed aromatic cluster model proposed by Yen (15) and a bridged aromatic model proposed by Speight (16) and by Strausz et al. (17) For a long time, asphaltene, the most polar and heaviest compounds of oil, has been believed to associate in solution to form complex colloidal structures. Gonzalez (*18*) studied asphaltene-resin interaction to point out that the descriptions representing asphaltenes as a solid phase, intrinsically insoluble in a hydrocarbon media dispersed by adsorbed resin molecules, does not accurately represent the structure of these fractions. Strausz et al. (*19*) reviewed the colloidal nature of asphaltenes in view of their covalent monomeric unit structure. Aggregation of asphaltenes is the cause of increased viscosity, emulsion stabilization, low solubility and eventually precipitation. Determination of the initial stage of aggregation has significant importance because it provides information regarding the concentration at which the aggregates are formed and how the interaction of asphaltenes leads to aggregates in the presence of other components of the oil matrix. Depending on the type of solvent, the measured Critical Micelle Concentration (CMC) for asphaltenes is about 2-18 g/L. Below CMC, asphaltenes should be in a molecular state, and above CMC asphaltenes form micelles. However, it is not clear whether below CMC the asphaltene aggregates are formed from asphaltene molecules or by smaller molecular associates such as dimers or trimers.

Recently Evdokimov et al. (*20*) investigated the initial stage of asphaltenes aggregation in dilute crude oil solutions using an NMR relaxation technique. The authors used dynamic viscosity measurements and spin-spin relaxation time T_2 from NMR studies, supplemented by optical absorption measurements, to elucidate the mechanism of asphaltene aggregate formation in crude oil-toluene solutions. It was found that the aggregation process of asphaltenes in a dilute solution of crude in toluene (100-150 mg/L) is a step process, changing from dimer to molecular nanoclusters (MNCs) of four stacked monomers. In asphaltene-containing solutions a major phase transition occurs (at $2 \times °C$) and at an asphaltene concentration of 150 mg/L due to molecular asphaltene interaction. From the experimental data it is not clear that the second phase transition occurs when asphaltene concentration is at "CMC" or if the phase transition occurs gradually. It has been suggested that the molecular self-association of asphaltenes in solutions is initiated at or above CMC. Asphaltenes molecules can associate through hydrogen bonding (*21*) or through charge transfer mechanism to form micelles. (*22*) A number of techniques have been applied for the observation of asphaltene self-association. The techniques used include surface tension, calorimetric, vapor pressure osmometry and small-angle neutron scattering. The break point in the results, for example, of surface tension measurements versus asphaltene concentration were interpreted as CMC values. Recently, near infrared (NIR) spectroscopy was applied for the determination of CMC. (*23*) In this work different concentration of asphaltenes in toluene were titrated with a non-polar solvent such as n-heptane and the onset of asphaltene precipitation was detected by NIR detector. A plot of (mL n-heptane/mL toluene) versus asphaltenes concentration yielded a curve with a break point that indicated the asphaltenes' CMC. The CMC values depended on the type of asphaltenes and the solvent used for the titration. It was further demonstrated that asphaltenes having different properties might have the same CMC.

Tanaka et al. (*24*) analyzed asphaltenes with small-angle neutron scattering to examine the changes in the structures of petroleum asphaltene aggregates in situ with solvents and temperature, and found a compact sphere the size of around 25Å in radius at 350 °C. Evdokimov et al. (*25*) also examined the assembly of asphaltene by near-UV/visible spectroscopy and found that the gradual aggregation process is distinct from conventional micellisation phenomena with step-like changes at CMC. They also concluded that asphaltene monomers were abundant in solutions with asphaltene concentrations below 1-5mg/L: the most stable oligomers were a dimer and a dimer pair. They also pointed out that these absorption results are supported by Rayleigh scattering results in asphaltene solutions.

It was also shown by Deo (*26*) that asphaltene precipitation onset points in toluene could be measured by titration by n-heptane and NIR spectroscopy. Aske et al. (*27*) used a high pressure NIR spectroscopy technique to determine the onset of asphaltene precipitation in live oils, and for asphaltenes dissolved in a mixture of toluene/pentane. The effect of pressure on redissolution of asphaltenes was also investigated. NIR in the electromagnetic spectrum ranged from 780 to 2500 nm (12820 to 4000 cm^{-1}). The attenuation of solutions of asphaltenes in the NIR path was both due to absorbance and scattering of light. Aggregation of asphaltenes as a result of depressurization resulted in baseline elevation of NIR. The fluid compressibility also caused base line elevation in high-pressure NIR. However, using Principal Component Analysis (PCA) compressibility and aggregation could be easily distinguished from each other. It was shown by using the combination of NIR with PCA that one could investigate the aggregation state of an asphaltene-containing system. Further, for the live oil using a NIR/PCA combination, it was shown that asphaltene aggregation is a reversible process.

In order to address a current controversy regarding the number of fused aromatic rings in the fused ring region core in petroleum asphaltenes, Ruiz-Morales et al. (*28*) studied the HOMO-LUMO gap as an index of molecular size and structure for polycyclic aromatic hydrocarbons and extended the findings to asphaltenes. Murgich et al. (*29*) obtained interesting results of the water effect on aggregation of asphaltenes: they performed molecular simulation of the association of petroleum asphaltenes, including association with water, and compared these with experimental results of heats of association obtained from titration calorimetry to verify that H bonding indeed may be an important factor in the association of asphaltenes. Water may act as a promoter of the association because of its small size and high intrinsic polarity.

There are several papers concerning molecular dynamic simulation of asphaltenes: Takanohashi (*30*) has reported the MDS of the heat induced relaxation of asphaltene aggregates, and Rogel (*31*) discussed the simulation of interaction in asphaltene aggregates such as the interaction between the resin fraction and asphaltene fraction. León (*32*) described the adsorption of alkyl-benzene-derived amphiphiles on an asphaltene surface using molecular dynamics simulation. Pacheco-Sanchez (*33*) reported the asphaltene aggregation under vacuum conditions at different temperatures by molecular dynamics to estimate that many asphaltene aggregates are not in stacking form at the end of every MD simulation and they are retained in irregular, jammed forms. There are at least three different geometric piling structures for asphaltene self-association, face to face geometric stacking, T-shaped(π-σ) and offset π stacked(σ-σ) geometric orientations.

Rogel (*34*) also applied MD simulation for the estimation of density of asphaltenes isolated from Venezuelan crude oils of different origins. The calculated values (1.01-1.15 g/cm^3) corresponded well to the experimental values (1.16-1.28 g/cm^3), then, the effect of different structural factors on caluculated densities of asphaltenes was systematically studied. Aufem et al., (*35*) in an interesting study on the interaction between asphaltenes and naphthenic acids, applied a pulsed field gradient-spin echo nuclear magnetic resonance technique for the determination of asphaltene particle sizes and found, in general, that traditional techniques based on light transmission or light scattering might suffer from the dark color of the sample. According to these methods, a dramatic decrease in diffusion coefficients of asphaltenes was observed upon increased concentration, indicating asphaltenes are prone to self-associate.

Mullins (*36*) has probed the order in asphaltene and aromatic ring systems by high resolution transmission electron microscopy. In Gray's report, mentioned earlier (*14*) it is very difficult to correlate the structure of asphaltenes to reactivity, however, he tried to do so. As for the chemical structure of asphaltenes, the RICO reaction, oxidation reaction can give us valuable information about the bridge bonds. Strausz et al. proposed an exciting structure for Asabasca asphaltenes based on their long-term well-organized research work. That structure needs some alteration in view of three dimensional arrangement, however, RICO reaction confirmed the presence of aliphatic bridge bonds in asphaltene structure. Carbognani and Rogel (*37*) recently had very interesting results on solvent swelling of petroleum asphaltenes and proposed a schematic structure swollen by the presence of non-polar solvent. In the pyrolysis process so many reactions take place that researchers have to conduct further fundamental study on the reaction using the model compounds to get detailed knowledge of the chemistry of asphaltenes.

Utilization

Stringent fuel specifications with decreasing crude oil quality increases the pressure on both process technology providers and refiners. Technology providers are implementing, and, at the same time, improving their existing key bottom-of-the-barrel process technologies, while refiners have been improving their operations to process more heavy and cheap crude oils. This, in turn, leads to increasing volumes and decreasing quality of residues. Residue is either blended with other low quality cutter stock (i.e., FCCU HCO, LCO) and sold as fuel oil or upgraded to high value products by utilizing either hydrogen addition processes (i.e., hydroprocessing) or carbon rejection processes (i.e., coking). A review of research and development activities showed that both academia and industry have been working to find advanced solutions in bottoms utilization. The reported literature was on reactive characterization, fundamental catalyst development, pyrolysis, hydroprocessing of whole crude oil or crude oil-derived residue.

Reactive Characterization

Asphalthenes obtained from Arabian heavy and medium crude oils were subjected to oxidation and alkylation reactions in an attempt to investigate the structural changes. (38) Alkylated (via Fidel-Crafts alkylation reactions with $AlCl_3$) and oxidized (with permanganate) asphaltenes were analyzed by Infrared NMR and thermogravimetric methods to observe the structural changes. Heavy ends separated from virgin, hydrocracked bitumen and coked using supercritical pentane extraction techniques were compared to pentane insoluble asphaltenes. (39) These heavy ends were analyzed/characterized by several analytical techniques including XPS, NMR, and GPC. It was reported that heavy ends obtained from supercritical pentane extraction were similar in terms of yields and properties to those obtained by traditional insolubility method. Heavy ends from virgin bitumen were comprised of 11 to 13 ring aromatic molecules. Similarly, hydrocracking pitch contained the highly aromatic heaviest molecules. It was suggested that removal of these heavy compounds prior to hydrocracking would be beneficial to the hydrocracking process. The chemical structural changes of asphaltenes were investigated during hydrotreating of Maya heavy crude oil using Vapor Pressure Osmometry (VPO) and NMR analysis. (40) Hydrotreating experiments were carried out in a pilot plant at temperatures ranging from 380-440 °C. While the nitrogen and metal content (Ni and V) of C_7 asphaltenes increased, the sulphur content decreased as a function of the reaction temperature. It was shown that at temperatures lower than 420 °C, dealkylation was a prominent reaction, whereas at higher temperatures hydrocracking of asphaltenes molecules led to lower molecular weight. The aromaticity of asphaltenes also increased as the temperature of the reaction increased. Since, during hydrocracking conversion, the protecting resin molecules are converted or destroyed, the asphaltene cores become more prone to precipitation/sludge formation and finally cause catalyst deactivation. Monitoring asphaltene stability during reaction is essential for process optimization.

New automated methods have been developed for measuring the flocculation onset of asphaltenes that have been or are in the process of being accepted as standard analytical techniques. (*41*) Sato et al. applied supercritical water oxidation to convert asphalt at temperatures in the range 340-400 °C and at water densities in the range 0-0.5 g/cm^3, under argon and air atmosphere. (*42*) Experiments were conducted in a micro tube reactor and asphaltene conversion as high as 47 Wt% was obtained at 400 °C under argon atmosphere. Desulfurization did not follow the cracking trend and was as high as 38 Wt% at 400 °C under air atmosphere. Results indicated that cracking reactions were favored under inert environment and desulfurization reactions were favorable under air. Asphaltene conversion increased with increasing water density. Athabasca bitumen-derived residues, fractionated using ASTM D1160 method, were further fractionated using a short-path distillation unit to produce three residue fractions boiling above 575 °C, 625 °C, 675 °C. (*43*) Fractions were analyzed and subjected to thermal hydrocracking in a semi-batch 18 mL autoclave at 400 °C, at 13.9 Mpa pressure for 30 minutes. Coking propensity was also studied using hot-stage microscopy at 400 °C and under 5.3 MPa hydrogen and nitrogen mixture. It was shown that as the residue cutpoint increased, contaminant levels, asphaltenes and molecular levels increased. Coke yield and asphaltene conversion also increased with increasing boiling point. Although asphaltene content increased with increased boiling point, maltenes were in sufficient quantities to keep them in solution. A coking propensity study showed that cut point and gas type, under which these experiments were conducted, did not impact the coke induction period.

Solid precipitation or fouling, which is an important phenomenon in handling and utilization of heavy oils, takes place when heavy oils are subjected to heat or blended with incompatible solvent or cracked. An apparatus was designed (*44*) to evaluate the temperature effect on the heat-induced deposition of five different heavy oils, namely Redwater B.C., CA Coastal, Boscan, MAXCL, and Vistar. Experiments were conducted at temperatures in the range 100-300 °C under Argon gas atmosphere, and solid deposition was determined by microscope. A good correlation was obtained with the solid deposition versus free solvent volume at high temperature (250 °C). The heat induced deposition was not important at low temperatures (100 °C) and started at 175 °C.

Coke deposition in a fractionator tower, resulting from the delayed coking process, was the subject of a paper by Gentzis and Rahimi. (*45*) Coke samples obtained from the commercial coking operations were analyzed by means of reflective light microscopy to determine coke formation mechanism. This study showed that coke precursors were entrained to the fractionator from the gas phase. The shape of the mesophase coke particles was attributed to the aromatic nature of the gas oil.

Catalyst development studies

Watanabe et al. (46) studied the thermal and dispersion behavior of oil soluble molybdenum dithiocarbamate and molybdenum dithiophospahte as MoS_2 precursors in petroleum residue derived from Arabian heavy crude oil. Among the spectroscopic methods used, FT-Far IR was shown to detect and monitor Mo complexes and their derived MoS_2 in vacuum residue. Molybdenum dithiocarbamate decomposed at 350 °C to form definite MoS_2 in vacuum residue, whereas molybdenum dithiophosphate started to decompose around 200 °C. However, no definite formation of MoS_2 was observed when the sample was heated to 500 °C.

Asphaltenes molecules, which contain high concentration of heteroatoms (i.e., Ni, V, S), play a key role in cracking reactions resulting in rapid catalyst deactivation. In an attempt to develop a better catalyst system for asphalthenes conversion, Inoue (47) studied the improvement of sepiolite-based catalysts for hydrodemetallization service. The objective was to increase the hydrodesulfurization activity of the sepiolite-based catalyst, which was reported to be better than alumina-based catalysts in asphaltenes cracking and metals removal. The sepiolite-based catalyst was improved by blending of alumina to the carrier and changing the amount of active metals. Hydrodesulfurization activity of sepiolite-based catalyst increased by about 40 % when the alumina content was increased from 25 Wt% to 50 Wt%. Hydrodemetallization reactivity remained the same when alumina contents were increased. In an explanatory study, (48) asphaltenes conversion was investigated over a Fe/SBA-15 catalyst in a mesoporous molecular sieve support, in a fixed-bed quartz reactor at 300 °C under atmospheric helium for 1 hour. Catalysts with four different pore sizes, ranging from 4.5 nm to 15 nm, were prepared and tested with asphaltenes derived from a Middle Eastern vacuum residue. The conversion increased with increasing pore size up to 12 nm and then leveled off. Pore structures were found to be stable after cracking and calcinations thereafter.

Pyrolysis

The effect of solvent type in asphaltene cracking was studied (49) to determine the solvent impact on reaction. Asphaltenes separated from the Athabasca vacuum residue mixed with solvents such as maltenes, 1-methynaphhalene, naphthalene, and tetralin were pyrolyzed in a 15 cc micro reactor at 430 °C under atmospheric pressure (in the case of maltenes) and at 3-4 MPa pressure (in the cases of other solvents used). It was concluded that the hydrogen-donating ability of solvents suppressed the coke formation. A kinetic model was developed for the phase separation.

Pyrolysis of heavy oils was the subject of several kinetic mechanism or catalyst development studies. Yasar et al. (*50*) studied the pyrolysis kinetics of saturates obtained from two Turkish crude oils, a paraffinic-based Luleburgaz crude oil and paraffinic and asphaltic-based Yenikoy crude oils. The objective was to gain insight on the heavy oil pyrolysis mechanism. Pyrolysis kinetics of saturates alone and in blend with asphaltenes was studied in tubular reactors at temperatures in the range 400-500 °C and residence times in the range of 30-300 min. First order pyrolysis kinetics showed that mixtures of asphaltenes and saturates are more reactive at 400 °C. The reactivities were, however, similar at 450 °C and 500 °C.

In a different application on the utilization of heavy oils, pyrolysis of bitumen was studied (*51*) in the presence and absence of water to evaluate chemical transformation of the bitumen matrix, which is used for inerting of homogeneous nuclear wastes in the deep geological repository. Autoclave experiments were conducted for whole bitumen and its fractions, namely: saturates, aromatics, resins and asphaltenes at temperatures in the range 200-400 °C at 5 MPa pressure and at a residence time of 72 hours. Above 300 °C, bitumen undergoes cracking and condensation reactions to produce light gas and insoluble residue. It was concluded that the sensibility of asphalts could be used for embedding of nuclear wastes to an elevation of temperature. Thermal cracking behavior of asphalthenes was also the subject of a study by Wang and Anthony. (*52*) The data from a previously published study was used to model the thermal cracking reactions to predict the behavior of asphalthenes at high temperatures. A three-component model, describing the reactions (asphaltenes to oil and gas, asphaltenes to coke, and oil and gas to coke) was used. Empirical results for coke formation and asphaltenes conversion gave a good fit to the experimental data. The model, which is not dependent on assumed reaction orders of cracking, represented the cracking behavior at high conversion level or long residence time. Thermal cracking of Cold Lake vacuum bottoms and their fractions, separated by column chromatography, were subjected to thermal hydrocracking reactions to investigate the effect of solvent and asphaltenes concentration on coke yield and to determine the reactivity of the asphaltene fractions. (*53*) Reactions were conducted at 440 °C under hydrogen pressure of 13.8 MPa in a batch autoclave. Asphaltene conversion was the highest when asphaltenes were reacted alone and lowest when they were reacted in solution in vacuum bottoms. Asphaltenes fractions, separated based on different solvent polarities, yielded fractions with increasing molecular weight. When hydrocracked, coke yield showed an increasing trend with increasing molecular weight. However, a similar trend was not observed for gas yields.

Hydrocarbon molecules undergo a change under steam stimulation conditions during the recovery process of heavy oil deposits. Hongfu et al. (*54*) investigated the molecular change of Liaohe heavy oils from China during the stream stimulation conditions in a glass tube, in which the oil and water mixture was heated to 240 °C for 24-72 hours. Three types of heavy oils from different fields were tested and molecular changes were monitored by SARA (Saturate, Aromatics, Resins and Asphaltenes) analysis using HPLC. Heavy molecules such as resins and asphaltenes were reduced in the range 20-30 Wt%, while aromatics and saturates fractions increased. The reduction of heavy fractions resulted in a substantial (as high as 42 %) viscosity reduction of heavy oils.

Apart from conventional processing schemes, Dunn and Yen (55) investigated the mechanism of asphaltene conversion with ultrasound at room temperature and pressures. The objective of the study was to investigate the transfer of hydrogen from a water source. Reactions were monitored by NMR and naphthenic hydrogen was found to be the most abundant hydrogen type. Dehydrogenation and cracking were the main reactions taking place. Hydrogen is the key component in reaction selectivity. In the absence of a hydrogen source, the dehydrogenation reactions were favored over cracking reactions (4:1 ratio). However, the selectivity shifted toward cracking (about 1:1) when the hydrogen source was available.

Hydroprocessing

In practice, an emphasis has been placed on processing bottom-of-the-barrel so as to reduce the low quality bottoms in refineries. In an attempt to increase the quality of heavy crude oils, hydroprocessability of Maya crude oil was been studied in a two-stage fixed-bed pilot plant at milder pressures (70 Kg/cm^2) than typical hydroprocessing pressures, and at 360-400 °C reaction temperatures and 0.5-2.0 h^{-1} liquid hourly space velocities. (56) Two types of catalysts were used: NiMo/Al$_2$O$_3$ in the first stage and CoMo/Al$_2$O$_3$ in the second. The quality of Maya crude oil increased substantially: API Gravity improved by about 8 degrees, while sulfur (78 Wt%), nickel and vanadium. (52.5 Wt%), and asphaltene content (62 Wt%) decreased substantially after two-stage hydroprocessing. Catalyst deactivation in a hydrodesulfurization process was studied (57) with Arabian light atmospheric residue to determine the effect of crude type on catalyst deactivation. Pilot plant tests were conducted separately: Hydrodemetallization experiments were conducted in a pilot plant utilizing 300 cc catalyst, and hydrodesulfurization reactions were conducted in a micro-flow unit utilizing 15 mL of stabilized catalyst and products collected from hydrodemetallization experiments used as feedstock. Asphaltene aromaticity, which is determined from ^{13}C NMR in liquid state (in CDCl$_3$), was found to be a good index for catalyst deactivation. Results with Kuwait and Arab light residue indicated that the fouling rate was a function of asphaltene aromaticity and hydrodesulfurization process temperature.

References

1. Alboudwarej, H.; Beck, J.; Svrcek, W.Y.; Yarranton, H.W.; Akbarzadeh, K. Sensitivity of asphaltene properties to separation techniques, *Energy Fuels* **2002**, *16*, 462.
2. Leontaritis, K. J.; Mansoori, G. A. Asphaltene deposition during oil production and processing: A thermodynamic colloidal model, *SPE International Symposium on Oilfield Chemistry*, San Antonio, TX, 1987, SPE 16258.
3. Acevedo, S.; Escobar, G.; Ranaudo, M.A.; Gutiérrez, L.B. Discotic shape of asphaltenes obtained from g.p.c. data, *Fuel* **1994**, *73*, 1807.
4. Gutiérrez, L.B.; Ranaudo, M.A.; Méndez, B.; Acevedo, S. Fractionation of asphaltene by complex formation with ρ-nitrophenol. A method for structural studies and stability of asphaltene colloids, *Energy Fuels* **2001**, *15*, 624.
5. León, O.; Contreras, E.; Rogel, E.; Dambakli, G.; Acevedo, S.; Carbognani, L.; Espidel, J., Adsorption of native resins on asphaltene particles: a correlation between adsorption and activity, *Langmuir* **2002**, *18*, 5106.
6. Cimino, R.; Correra, S.; Del Bianco, A.; Lockhart, T.P. In *Asphaltenes: Fundamental and Application*; Sheu, E.Y., Mullins, O.C.; Eds.; Plenum Press: New York, 1995; p. 97.
7. Buckley, J.S.; Hirasaki, G.J.; Liu, Y.; Von Drasek, S.; Wang, J.X.; Gill, B.S. Asphaltenes precipitation and solvent properties of crude oils, *Pet. Sci. Technol.* **1998**, *16* (3&4), 251.
8. Buckley, J.S. Predicting the onset of asphaltene precipitation from refractive index measurements. *Energy Fuels* **1999**, *13*, 328.
9. Wang, J.; Buckley, J. Asphaltene stability in crude oil and aromatic solvents -- the influence of oil composition, *Energy Fuels* **2003**, *17*, 1445.
10. Oh, K.; Deo, M.D. Effect of organic additives on the onset of asphaltene precipitation, *Energy Fuels*, **2002**, *16*, 694.
11. Feng Hu, Y.; Li, S.; Liu, N.; Ping Chu, Y.; Park, S.; Ali Mansoori, G.; Min Guo, T. Measurement and corresponding states modeling of asphaltene precipitation in Jilin reservoir oils, *J. Petrol. Sci. Eng.* **2004**, *41*, 169.
12. Idem, R.; Ibrahim, H. Kinetics of CO_2-induced asphaltene precipitation from various Saskatchewan crude oils during CO_2 miscible flooding, *J. Petrol. Sci. Eng.* **2002**, *35*, 233.
13. Vafaie-Sefti, M.; Mousavi-Dehghani, M.; Mohammad-Zadeh, M. A simple model for asphaltene deposition in petroleum mixtures, *Fluid Phase Equilibria* **2003**, *206*, 1.
14. Gray, M.R. Consistency of asphaltene chemical structures with pyrolysis and coking behavior, *Energy Fuels* **2003**, *17*(6), 1566.
15. Yen, T.F. *Prepr. –Am. Chem. Soc., Div. Pet. Chem.* 1972, 17(4), F102.
16. Speight, J.G. *The chemistry and technology of petroleum*, 2nd ed.; Marcel Dekker: New York, 1991.

17. Murgich, J.; Abanero, J.A.; Strausz, O.P. Molecular recognition in aggregates formed by asphaltene and resin molecules from the Athabasca oil sand, *Energy Fuels* **1999**, *13*, 278.
18. González, G.; Neves, G.B.M.; Saraiva, S.M.; Lucas, E.F.; dos Anjos de Sousa, M. Electrokinetic Characterization of asphaltenes and the asphaltenes--resins interaction, *Energy Fuels*, **2003**, *17*, 879.
19. Strausz, O.P.; Peng, P.; Murgich, J. About the colloidal nature of asphaltenes and the MW of covalent monomeric units, *Energy Fuels* **2002**, *16*, 809.
20. Evdokimov, I.; Eliseev, N.; Akhmetov, B. Initial stages of asphaltene aggregation in dilute crude oil solutions: studies of viscosity and NMR relaxation, *Fuel* **2003**, *82*, 817.
21. Ho, B.; Briggs, D.E. Small angle X-ray scattering from coal-derived liquids, *Colloids Surf.* **1982**, *4*, 271.
22. Shue, E.Y.; Storm, D.A. In *Asphaltenes: Fundamental and Applications*; Shue, E.Y.; Mullins, O.C.; Eds; Plenum Press: New York, 1995; Chapter 1, p. 1.
23. Oh, K.; Oblad, S.C.; Hanson, F.V.; Deo, M.D. Examination of asphaltenes precipitation and self-aggregation, *Energy Fuels* **2003**, *17*, 508.
24. Tanaka R.; Hunt, J.E.; Winans, R. E.; Thiyagarajan, P.; Sato, S.; Takanohashi, T. Aggregates structure analysis of petroleum asphaltenes with small-angle neutron scattering, *Energy Fuels* **2003**, *17*, 127.
25. Evdokimov, I.N.; Eliseev, N.Y.; Akhmetov, B.R. Assembly of asphaltene molecular aggregates as studied by near-UV/visible spectroscopy I. Structure of the absorbance spectrum, *J. Petrol. Sci. Eng.* **2003**, *37*, 135.
26. Oh, K.; Oblad, S.C.; Hanson, F.V.; Deo, M.D. Examination of asphaltenes precipitation and self-aggregation, *Energy Fuels* **2003** *17*, 508.
27. Aske, N.; Kallevik, H.; Johnsen, E.E.; Sjöblom, J. Asphaltene aggregation from crude oils and model systems studied by high-pressure NIR spectroscopy, *Energy Fuels* **2002**, *16*, 1287.
28. Ruiz-Morales, Y. HOMO-LUMO gap as an index of molecular size and structure for polycyclic aromatic hydrocarbons (PAHs) and asphaltenes: A theoretical study. I, *J. Phys. Chem. A* **2002**, *106*, 11283.
29. Murgich, J.; Galeana, C. L.; Manuel del Rio, J.; Merino-Garcia, D.; Andersen, S. I. Molecular mechanics and microcalorimetric investigations of the effects of molecular water on the aggregation of asphaltenes in solutions, *Langmuir* **2002**, *18*, 9080.
30. Takanohashi, T.; Sato, S.; Saito, I.; Tanaka, R. Molecular dynamics simulation of the heat-induced relaxation of asphaltene aggregates, *Energy Fuels* **2003**, *17*, 135.
31. Rogel, E. Simulation of interactions in asphaltene aggregates, *Energy Fuels* **2003**, *14*, 566.
32. Rogel, E.; León, O. Study of the adsorption of alkyl-benzene-derived amphiphiles on an asphaltene surface using molecular dynamics simulations, *Energy Fuels*, **2001**, *15*, 1077.

16

33. Pacheco-Sanchez, J. H.; Zaragoza, I. P.; Martinez-Magadan, J. M. Asphaltene aggregation under vacuum at different temperatures by molecular dynamics, *Energy Fuels* **2003**, *17*, 1346.

34. Rogel, E.; Carbognani, L. Density estimation of asphaltenes using molecular dynamics simulations, *Energy Fuels* **2003**, *17*, 378.

35. Östlund, J.-A.; Nydén, M.; Auflem, I. H.; Sjöblom, J. Interactions between asphaltenes and naphthenic acids, *Energy Fuels*, **2003**, *17*, 113.

36. Groenzin, H.; Mullins, O.C.; Eser, S.; Mathews, J.; Yang, M.-G.; Jones, D. Molecular size of asphaltene solubility fractions, *Energy Fuels* **2003**, *17*, 498.

37. Carbognani, L.; Rogel, E. Solvent swelling of petroleum asphaltenes, *Energy Fuels* **2002**, *16*, 1348.

38. Siddiqui, M. Alkylation and oxidation reactions of Arabian asphaltenes, *Fuel* **2003**, *82*, 1323.

39. Zhao, S.; Kotlyar, L.; Woods, J.; Sparks, B.; Hardacre, K.; Chung, K. Molecular transformation of Athabasca bitumen end-cuts during coking and hydrocracking, *Fuel* **2001**, *80*, 1155.

40. Ancheyta, J.; Centeno, G.; Trejo, F.; Marroquin, G. Changes in asphaltene properties during hydrotreating of heavy crudes, *Energy Fuels* **2003**, *17*, 1233.

41. Vilhunen, J.; Quignard, A.; Pilviö, O.; Waldvogel, J. Use of the PORLA automated heavy fuel stability analyzer for the monitoring of the H-Oil plants, *LASH 2000, the 7ᵗʰ International Conference on Stability and Handling of Liquid Fuels*, Graz, Austria, 2000.

42. Sato, T.; Adschiri, T.; Arai, K.; Rempel, G.; Ng. F. Upgrading of asphalt with and without partial oxidation in supercritical water, *Fuel* **2003**, *82*, 1231.

43. Rahimi, P.; Gentzis, T.; Taylor, E.; Carson, D.; Nowlan, V.; Cotte, E. The impact of cut point on the processability of Athabasca bitumen, *Fuel* **2001**, *80*, 1147.

44. Schabron, J.; Pauli, A.; Rovani Jr, J.F. Non-pyrolytic heat induced deposition from heavy oils, *Fuel* **2001**, *80*, 919.

45. Gentzis, T.; Rahimi, P. A microscopic approach to determine the origin and mechanism of coke formation in fractionation towers, *Fuel* **2003**, *82*, 1531.

46. Watanabe, I.; Otake, M.; Yoshimoto, M.; Sakanishi, K.; Korai, Y.; Mochida, I. Behaviors of oil-soluble molybdenum complexes to form very fine MoS₂ particles in vacuum residue, *Fuel* **2002**, *81*, 1515.

47. Inoue, S.; Takatsuka, T.; Wada, Y.; Nakata, S. Ono, T. A new concept for catalysts of asphaltene conversion, *Catal. Today* **1998**, *43*, 225.

48. Byambajav, E.; Ohtsuka, Y. Cracking behavior of asphaltene in the presence of iron catalysts supported on mesoporous molecular sieve with different pore diameters, *Fuel* **2003**, *82*, 1571.

49. Rahmani, S.; McCaffrey, W.; Gray, M. Kinetics of solvent interactions with asphaltenes during coke formation, *Energy Fuels* **2002**, *16*, 148.

50. Yasar, M.; Cerci, F.; Gulensoy, H. Effect of asphaltenes on pyrolysis kinetics of saturates, *J. Anal. Appl. Pyr.* **2000**, *56*, 219.
51. Schlepp, L.; Elie, M.; Landais, P.; Romero, M. Pyrolysis of asphalt in the presence and absence of water, *Fuel. Process. Technol.* **2001**, *74*, 107.
52. Wang, J.; Anthony, E. A study of thermal-cracking behavior of asphaltenes, *Chem. Eng. Sci.* **2003**, *58*, 157.
53. Rahimi, P.; Gentzis, T. Thermal hydrocracking of Cold Lake vacuum bottoms asphaltenes and their subcomponents, *Fuel. Process. Technol.* **2003**, *80*, 69.
54. Hongfu, F.; Yongjian, L.; Liying, Z. Xiaofei, Z. The study on composition changes of heavy oils during steam stimulation processes, *Fuel* **2002**, *81*, 1733.
55. Dunn, K.; Yen, T. A plausible reaction pathway of asphaltene under ultrasound, *Fuel. Process. Technol.* **2001**, *73*, 59.
56. Ancheyta, G.; Betancourt, G.; Marroquín, G.; Centeno, G.; Castañeda, L.C.; Alonso, F.; Muñoz, J.A.; Gómez, M.; Rayo, P. Hydroprocessing of Maya heavy crude oil in two reaction stages, *Appl. Catal. A* **2002**, *233*, 159.
57. Seki, H.; Yoshimoto, M. Deactivation of HDS catalyst in two-stage RDS process II. Effect of crude oil and deactivation mechanism, *Fuel. Process. Technol.* **2001**, *69*, 229.

Chapter 2

Characterization of Constituents of Arabian Vacuum Residues by FT-ICR MS Coupled with Various Ionization Techniques

Keiko Miyabayashi, Yasuhide Naito, Kazuo Tsujimoto, and Mikio Miyake

School of Materials Science, Japan Advanced Institute of Science and Technology (JAIST), 1–1 Asahidai, Tatsunokuchi, Ishikawa 923–1292, Japan

Based on ultrahigh resolution mass spectra obtained by Fourier transform ion cyclotron resonance mass spectrometer (FT-ICR MS), chemical formula of every constituent in vacuum residue were determined directly without any pre-separation procedures. This study aims to evaluate features of ionization techniques combined with FT-ICR MS. When electrospray ionzation (ESI) was applied to Arabian mix vacuum residue (AM-VR) in methanol/chloroform, nitrogen-containing hydrocarbons, adducted with H^+ or Na^+, were detected as major peaks, while aromatic hydrocarbons, adducted with H^+ or Na^+, were detected as odd peaks by using ESI in methanol/ chloroform/trifluoroacetic acid (TFA). Hydrocarbons and sulfur-containing aromatic hydrocarbons were detected in similar intensities by electron ionization (EI). Liquid secondary ionization (LSI) detects hydrocarbons with and without nitrogen- and sulfur-atom. Thus, selection of proper ionization technique is very important to analyze specific types as well as variety of compounds in vacuum residues.

Mass Spectrometry on Heavy Oil

Heavy oil is an extremely complex mixture consisting of variety of compounds, which composition varies significantly depending on the production places and the refinery conditions. The averaged structural characteristics of heavy oil have been investigated by means of several analytical techniques, such as $^1H/^{13}C$-NMR (1,2), IR (3), X-ray (4) and the gel permeation chromatography (GPC). Although such average structures are very important to understand the properties of heavy oil, there is a possibility that minor constituents, not reflected to average structure, may play serious role during its processing, *e.g.*, catalyst deactivation (5).

Mass spectrometer (MS) is a viable tool to provide a compositional inventory of vacuum residue that is made up of thousands of compounds. Magnetic sector mass spectrometers coupled with several ionization sources have been applied to petroleum related materials. Constituents of Arabian vacuum residue are characterized by using field ionization mass spectrometer (FIMS) after separation by two-stage HPLC (6). A thermospray-interface (TSP) MS was applied to Middle East vacuum residue to evaluated the distributions of aromatic hydrocarbons (7). Matrix assisted laser desorption/ionization (MALDI) was used to determine the molecular weight distribution for asphaltene of petroleum vacuum residue (8,9). However, vacuum residue samples must be separated into several fractions by chromatography prior to analyze the chemical structure of each constituent because of insufficient resolution of conventional mass spectrometers (6).

Fourier transform ion cyclotron resonance mass spectrometer (FT-ICR MS), a new type mass spectrometer, has an ultra-high resolving power with a mass accuracy of less than 1 ppm (10^{-6}) (10). FT-ICR MS can be a viable method to directly analyze chemical structures of each constituent molecule in a very complex mixture by accurate mass measurements, when singly-charged molecular ions are preferentially produced (10). Several research groups have recently applied FT-ICR MS to oil related materials. FT-ICR MS equipped with 3T magnet and low-voltage electron ionization (LVEI) was applied to determine the molecular formulas of aromatic compounds in petroleum distillates (11,12). Rodgers *et al.* developed 5.6T FT-ICR MS connected to an all-glass heated inlet system with LVEI, and identified 500 aromatic compounds either with or without sulfur-atoms in diesel fuels (13). Qian *et al.* characterized nitrogen-containing aromatic compounds in heavy petroleum crude oil by the application of electro-spray ionization (ESI) FT-ICR MS (14).

ESI is known to be the softest ionization method avoiding fragmentation. At present, ESI FT-ICR MS is earning a great success for the structural analysis of bio-related materials, such as peptides and proteins (15,16). On the other hand, LVEI is proved to ionize non-polar aliphatic compounds in vacuum residues, which cannot be ionized by ESI, although fragmentation causes difficulty in spectral analysis of complex mixtures (17). The liquid secondary ionization (LSI) is known to have extensive ionization ability irrespective of polar and non-polar compounds without significant fragmentation.

In the present study, the results of FT-ICR mass measurements of Arabian vacuum residues obtained by three different ionization techniques, *i.e.*, ESI, EI and LSI, are discussed to characterize features of respective ionization methods. A selection of proper ionization technique is very important to analyze specific types as well as variety of compounds in vacuum residues.

Instrumentation of FT-ICR MS

Principle

The theory of FT-ICR mass spectrometry is based on detection of ion motion in magnetic field (10). Ions, present in vacuum and in spatially uniform magnetic field, rotate around magnetic field line affecting by Lorentz force of magnetic field. The constant angular velocity of the ion motion is the "unperturbed" cyclotron angular frequency, which is proportional to the magnetic flux density and is inversely proportional to mass-to-charge ratio m/z. Hence using a well-defined magnetic field density, mass-to-charge ratio can be determined by measuring the cyclotron resonance frequency. In FT-ICR mass spectrometry, an observed time domain data set is transformed into a frequency domain data one by Fourier transformation.

Instrument

In the present study, the FT-ICR mass spectra were recorded on a Bruker BioAPEXII 70e FT-ICR mass spectrometer (Bruker Daltonics Inc., Billerica, MA) equipped with a 7 T superconducting magnet. The FT-ICR MS experimental event sequences and parameters were adjusted to optimize sensitivity by the Bruker's XMASS software running on a Silicon Graphics INDY (R5000) computer system. Accumulating ions in the external ion reservoir (hexapole) for 3 s were transported through electrostatic lens systems to the ICR cell where mass separation and detection are conducted with excitation, detection, and trap electrodes. A broad band chirp excitation was used for all the experiments. Time domain data sets in the length of 256 k words were acquired using broad-band detection for all the mass spectra. Absolute peak intensity was obtained by using the peak pick module available in the Bruker's XMASS package. Each molecular formula (a possible combination of atoms, which total mass give the least deviation from the measured mass) was determined by using *mass analysis* module available in the Bruker's XMASS package. Details are described elsewhere (18,19).

FT-ICR MS with Electrospray Ionization (ESI)

Methodology and Features of ESI

ESI has been developed to apply biological macromolecules and polymers as the softest ionization method avoiding fragmentation and to ionize non-volatile materials on the basis of the adduct formation at polar functional groups (e.g., -OH, -NH$_2$) with H$^+$ or Na$^+$ contained in a solvent (20). The ion formation process is briefly described: Charged droplets are produced by spraying the sample solution through a needle, to which a potential about few kV are applied. By the evaporation of solvent from the droplets (*i.e.*, decrease in droplet size), Coulombic repulsion overcomes the droplet's surface tension, and then the droplet explodes to form sample ions. ESI usually forms adducted ion with polar functional groups; not suitable to hydrocarbons without polar functional groups, which are major constituents of heavy oil. For this reason, few applications of ESI MS have been performed in the characterization of petroleum-related materials.

Experimental

A vacuum residue sample was prepared from Arabian crude mixture (80% light and 20% medium) (AM-VR) and was provided by Nippon Oil Ltd. Co., which elemental composition (wt%) was as follows: C 84.3, H 9.9, S 5.2, N 0.4, and O 0.2 (diff.) (2). AM-VR sample of 1.2 mg was dissolved in the ESI solvent (1 ml): mixture of methanol/chloroform (0.8/0.2, v/v) or methanol/chloroform/ trifluoroaceteic acids (TFA) (1/0.1/0.01, v/v) and the solution was infused into the ESI source in a positive ion mode. The desolvation was accelerated by counter-current nitrogen gas heated at 250°C. A voltage of −3.5 kV was applied between the grounded needle and the metal-capped glass capillary. Detailed conditions were described previously (18,19).

ESI FT-ICR MS of AM-VR in Neutral Solvent

Figure 1 shows a broad band ESI FT ICR mass spectrum of AM-VR measured in methanol/chloroform. Many peaks were observed in the range of $250 < m/z < 750$. The spectrum in the range of $456 < m/z < 461$ is expanded in Figure 2 for the peak pattern analysis. The mass differences between pair of peaks with odd and even masses are $1.005 \sim 1.002$ Da, which coincide with the atomic weight difference between ^{13}C and ^{12}C (1.003 Da). The intensity ratios of peaks with odd/even masses are $30 \sim 45\%$, corresponding to estimated peak intensity ratios with odd/even masses according to the isotope existence ratio of ^{13}C (1.11% of C) and the plausible carbon number of ca. 30 for molecule with around 450 Da. Therefore the peaks with odd masses correspond the compounds including one ^{13}C. A series of peaks with even masses appeared at intervals $2.017 \sim 2.015$ Da which correspond to 2.016 Da of 2H (Figure 2). Another sequence of peaks was detected with a mass difference of 14.016 Da (CH_2). Essentially a similar spectral pattern was obtained for all the other peaks observed in the range of $350 < m/z < 800$. The molecular formulas for the even m/z peaks can be determined based on highly accurate masses. Molecular formulas for the even m/z peaks can be determined by the calculation of formula mass with the least deviation to the measured mass. For the formula mass calculation, it was assumed that molecules were composed of ^{1}H, ^{12}C, ^{16}O (less than 2), ^{14}N (less than 2) and ^{32}S (less than 2). Table I gives an example of the candidate molecular formulas for the peak observed at m/z 466.3460 in Figure1. The molecular formula $C_{34}H_{44}N$ (formula mass 466.3468 Da) is the most probable one because it has the least deviation (0.8 mDa) from the measured m/z 466.3460. Qian *et al.* also detected nitrogen-containing molecules as protonated ions in crude oil by micro ESI FT-ICR MS with the solvent system of methanol/methylene chloride/acetic acid (21). Figure 3 summarizes results of the molecular formula analyses for the all peaks with even masses observed in the range of $570 < m/z < 670$ in Figure 1. The peak intensities are plotted against m/z value and molecules having the same carbon number are linked with a dashed-line. Z-values (hydrogen deficiency index, defined as $[C_nH_{2n+z}N_mS_s+H]^+$) are denoted in the diagram as a measure of degree of aromatic ring condensation. One nitrogen-containing hydrocarbons with carbon numbers from 41 to 49 were detected, where the Z-values varied from -47 to -9. Interestingly, the maximum intensity is observed at the Z-value of around -25, irrespective of either the carbon numbers or the m/z values. This result suggests that a series of detected molecules in AM-VR may have a similar condensed aromatic-ring size with different lengths of alkyl side-chains.

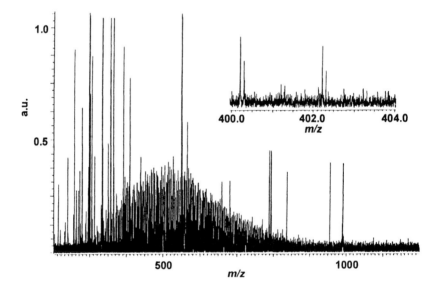

Figure 1. ESI FT-ICR mass spectrum of AM-VR in methanol/chloroform. Inset shows the expanded view.

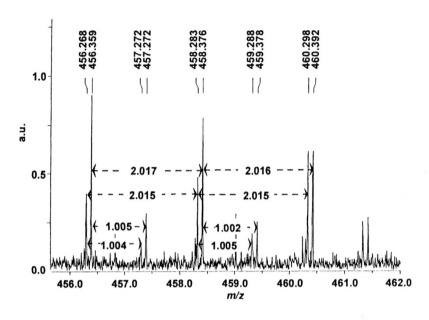

Figure 2. Mass scale-expanded view of signals near m/z 458 in Figure 1.

Table I . Candidate Molecular Formulas Having Close Masses to a Peak with *m/z* 466.3460

Formula mass / Da	Difference / mDa	Z*	Molecular formula				
			^{12}C	^{1}H	^{14}N	^{32}S	^{16}O
466.3468	0.8	-25	34	44	1	–	–
466.3502	4.2	-15	31	48	1	1	–
466.3410	-5.0	-3	27	52	2	2	–
466.3536	7.6	-3	28	54	1	2	–
466.3376	-8.4	-15	30	46	2	1	–
466.3554	9.4	-15	30	46	2	–	2

*Z-value : $[C_nH_{2n+Z}N_mS_sO_o+H]^+$

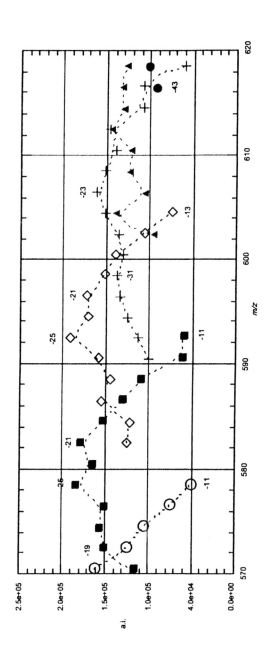

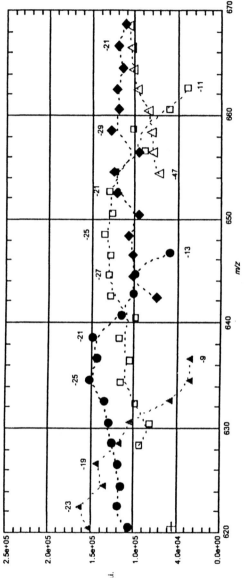

Figure 3. Relation between intensity of even peak and m/z. Numerical values denote Z-value $[C_nH_{2n+Z}N+H]^+$. Molecules having the same carbon number are linked with a dashed line; C_{41} (○), C_{42} (■), C_{43} (◇), C_{44} (+), C_{45} (▲), C_{46} (●), C_{47} (□), C_{48} (◆), and C_{49} (△).

ESI FT-ICR MS of AM-VR in Acidic Solvent

As described above, nitrogen-containing compounds are detected as major peaks when neutral solvent (chloroform/methanol) was used for ESI FT-ICR mass measurement. It is well known that major constituent in heavy oil is hydrocarbons and that nitrogen-containing compounds are minor ones. Then, we have investigated ESI solvent composition to detect hydrocarbons. We examined effect of strong acid, such as trifluoroaceteic acid (TFA), to promote protonation of aromatic hydrocarbon constituents in heavy oil. ESI FT-ICR mass spectrum of AM-VR by using methanol/chloroform/TFA is shown in Figure 4. The peaks distributed over m/z 1000 in methanol/chloroform/TFA; higher m/z region than that in methanol/chloroform (see Figure 1). Molecular formulas for the several odd m/z peaks in Figure 4 are determined by the accurate mass values, in a manner similar to that described in Table I. Hydrocarbons are detected as H⁺ adducted ions (Table II). When coronene was used as a model compound in ESI FT-ICR mass measurements, the protonated ion was observed as primarily peaks in methanol/chloroform/TFA, consistent with the results of AM-VR spectra. Thus, tunable effect of detected compounds has been demonstrated as even and odd peaks for one nitrogen-containing compounds and pure hydrocarbons, respectively, by changing ESI solvent compositions.

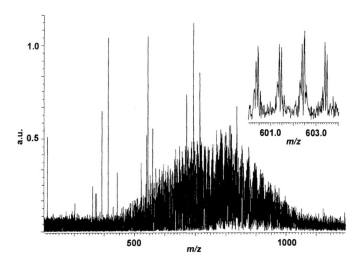

Figure 4. ESI FT-ICR mass spectra of AM-VR in methanol/chloroform/TFA. Inset shows the expanded view.

Table II. Estimated Molecular Formulas for Observed Peaks with *m/z* 601.3828, 603.4012 and 649.4743 in Figure 4

Measd. Mass / Da	Molecular formula	Formula mass / Da (Difference / mDa)	Z*
601.3828	$[C_{46}H_{48}+H]^+$	601.3829 (-0.1)	-44
603.4012	$[C_{46}H_{50}+H]^+$	603.3985 (-2.7)	-42
649.4743	$[C_{49}H_{60}+H]^+$	649.4767 (-2.4)	-38

*Z-value : $[C_nH_{2n+Z}+H]^+$

FT-ICR MS with In-beam Electron Ionization (In-beam EI)

Methodology and Features of In-beam EI

EI is the most widely adopted ionization technique, which can ionize variety of compounds so long as the compound is vaporized. Application of EI to low-volatile sample causes fragmentation due of thermal decomposition during vaporization. Samples are introduced as vapor into the path of the electron beam, thermally emitted from a tungsten filament. A permanent magnet is positioned across an ion chamber to produce a magnetic flux in parallel to the electron beam. This causes the electron beam to spiral from the filament to the trap, resulting in increase the ionization efficiency.

In-beam EI is applied to ionize rather low-volatile sample with large molecular weight, where a sample places close to electron beam path to increase ionization efficiency (17). Typically platinum wire to put a sample is inserted into a platinum sample vial to shorten the distance from a filament.

Experimental

Arabian Heavy vacuum residue (AH-VR) was adopted as a sample. The maltene, soluble fraction of AH-VR in *n*-heptane, was subsequently separated by alumina column chromatography and the fractions eluted by *n*-heptane, toluene, and toluene/methanol/chloroform were defined as AH-VR-Sa, AH-VR-Ar and AH-VR-Re, respectively. The AH-VR-Sa sample was subjected to in-beam EI FT-ICR mass measurements.

A commercial probe was modified for in-beam EI by inserting a platinum wire (15 mm in length, 1.1 mm in diameter) into a platinum vial. The distance from the end of the platinum wire to the filament is around 1 cm. The wire can be heated up to 350 °C. The probe temperature and the ionization energy were independently changed from 150 to 350 °C and from 20 to 70 eV. The extraction voltage of ion source was set to −6.6 V. The ionization pulse length was set to 50 ms. A source housing temperature was maintained at 150 °C to assist vaporization of the samples.

In sample preparation for mass measurements, the each fraction (1 mg) was dissolved in 1 ml of methylene chloride. 0.5 μl of the each sample solution was loaded on the in-beam EI probe tip. Details are described elsewhere (22).

In-beam EI FT-ICR MS of AH-VR-Sa

When ESI was applied to AH-VR-Sa, no peak was obtained by FT-ICR MS, presumably because of aliphatic and non-polar nature of the sample. Thus, an another ionization method, *i.e.*, in-beam EI, is investigated to detect components in the AH-VR-Sa.

Figure 5 shows in-beam EI FT-ICR mass spectrum of AH-VR-Sa obtained under the ionization energy and probe temperature at 30 eV and 300°C, respectively. These conditions were selected after optimization to suppress fragmentation and to detect as much as peaks. All spectra were detected in the positive ion mode. The peaks are found in the range of $100 < m/z < 650$. Mass scale expanded segment derived from Figure 5 is shown in Figure 6. The peak intensity pattern of the in-beam EI FT-ICR mass spectrum is significantly different from those of ESI FT-ICR mass spectrum shown in Figure 2. Estimated molecular formulas based on accurate masses are also shown in Figure 6. The pair of peaks consists of hydrocarbons with and without one sulfur-atom. The chemical formulas of all the hydrocarbons detected in the range of $100 < m/z < 630$ are determined. Hydrocarbons with low m/z (C-number) region consist of high Z-number, suggesting rather small molecules are rich in aliphatic carbons. On the other hand, hydrocarbons with low Z-values concentrate in high m/z (C-number) region, denoting that large molecules are consisting of developed aromatic rings. The chemical formulas of all the sulfur-containing hydrocarbons detected in the range of $80 < m/z < 570$ are also determined (Figure 7). The maximum abundance appears irrespectively of carbon number at a similar Z-value of −10, which corresponding to benzothiophene derivatives.

Therefore, significantly different types of compounds were detected by using EI compared with those by ESI.

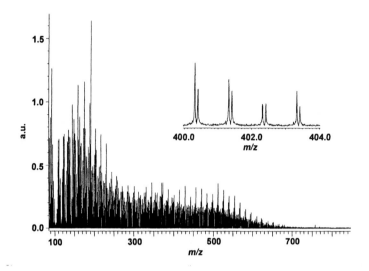

Figure 5. FT-ICR mass spectrum of AH-VR-Sa obtained in the optimized in-beam EI conditions; the probe temperature and the ionization energy were 300 °C and 30eV, respectively.

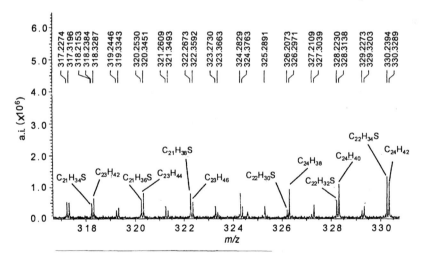

Figure 6. Mass scale-expanded view of signals in Figure 5.

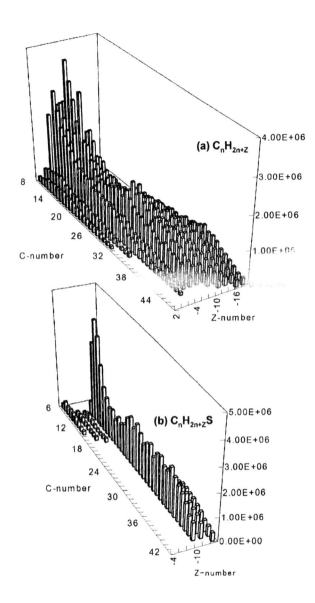

Figure 7. Compounds type distributions observed in in-beam EI FT-ICR mass spectrum; (a) C_nH_{2n+Z} and (b) $C_nH_{2n+Z}S$.

FT-ICR MS with Liquid Secondary Ionization (LSI)

Methodology and Features of LSI

LSI is comparatively soft ionization technique and is well suited to a low volatile sample, typically producing adducted-molecular ion species denoted as $[M+H]^+$ and $[M-H]^-$ (23). The particle beam, typically Cs^+, is irradiated to a sample mixture with a matrix (e.g., glycerol or 3-nitrobenzyl alcohol). The species, ejected from the surface as secondary ions by this bombardment, are extracted from the source and analyzed by the mass spectrometer.

Experimental

AM-VR sample was mixed with liquid matrix (3-nitrobenzyl alcohol). About 1 μL of the mixed slurry was directly loaded on the LSI probe tip. The primary ion (Cs^+) gun was operated at an anode potential of 10 kV. The extraction voltage in the source was 10 V and the ionization pulse length was set to 600 ms. The measured spectrum was internally calibrated by using matrix cluster ions. FT-ICR operation was the same as ESI described above. All spectra were detected in the positive ion mode.

LSI FT-ICR MS of AM-VR

FT-ICR mass spectrum obtained by using LSI is shown Figure 8. A rather low mass range detected is presumably due to the low acceleration voltage (10 kV) of primary Cs^+ ion. The mass scale expanded segment of AM-VR LSI mass spectrum is shown in Figure 8 inset. Although peak pattern were rather similar to EI shown in Figure 6, every peak with an odd mass was more abundant than that with an adjacent even mass over the whole spectrum region. The results of molecular formula analysis for several peaks based on accurate mass are shown in Table III. Hydrocarbons, compounds with one sulfur-atom and compounds with one nitrogen-atom ($[C_nH_{2n+Z}+H]^+$, $[C_nH_{2n+Z}S+H]^+$, and $[C_nH_{2n+Z}N+H]^+$, respectively) are observed as protonated ions. The deviation between measured and calculated masses is less than 1.1 mDa.

The distribution of compounds estimated for all the peaks in the LSI mass spectrum in the range of $230 < m/z < 330$ are summarized in Figure 9. The most abundant constituents were one sulfur-containing compounds with carbon and Z-values of 17~18 and −18~−22, respectively. Considerable difference in compound type distribution was observed for hydrocarbons, sulfur-containing and nitrogen-containing compounds.

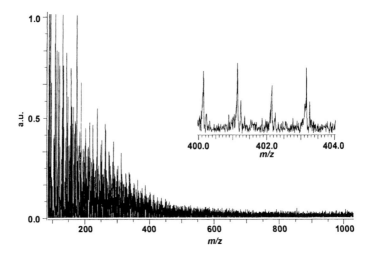

Figure 8. LSI FT-ICR mass spectrum of AM-VR. Inset shows expanded view.

Table III. Estimated Molecular Formulas for the peaks *m/z* 271.1503, 272.1426. 274.1588, 275.0900 and 275.1796 in Figure 8

Measd. mass / Da	Molecular formula	Formula mass / Da (Difference / mDa)	Z*
271.1503	$[C_{18}H_{22}S+H]^+$	271.1515 (0.8)	-14
272.1426	$[C_{20}H_{17}N+H]^+$	272.1434 (-0.8)	-23
274.1588	$[C_{20}H_{19}N+H]^+$	274.1590 (0.2)	-21
275.0900	$[C_{19}H_{15}S+H]^+$	275.0899 (1.1)	-23
275.1796	$[C_{21}H_{22}+H]^+$	275.1794 (0.2)	-20

*Z-value : $[C_nH_{2n+Z}N_mS_s+H]^+$

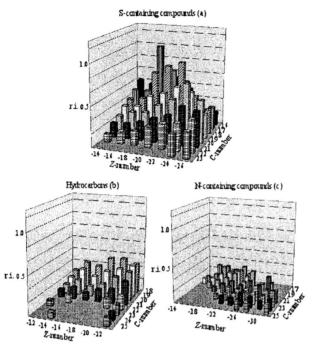

Figure 9. The Compounds distribution detected in LSI FT-ICR mass spectrum of AM-VR; (a) S-containing compounds, (b) Hydrocarbons, and (c) N-containing compounds.

Conclusion

Fourier transform ion cyclotron resonance mass spectrometer (FT-ICR MS) with 7 T superconducting magnet, which attains ultrahigh resolution, can provide chemical formula of every constituent in Arabian vacuum residue samples directly without any pre-separation procedures based on accurate mass measurements and spectral pattern analyses. The characteristics of ionization techniques combined with FT-ICR MS are summarized as follows.

1. When electrospray ionization (ESI) was applied to Arabian mix vacuum residue (AM-VR) in methanol/chloroform (neutral solution), nitrogen-containing hydrocarbons, adducted with H^+ or Na^+ ($[C_nH_{2n+z}N_mS_s+H]^+$ or $[C_nH_{2n+z}N_mS_s+Na]^+$) , were detected as major peaks in the range of $250 < m/z < 750$. Chemical formulas analyses of all even peaks observed in the range of $570 < m/z < 670$ indicate that one nitrogen-containing hydrocarbons with carbon numbers from 41 to 49 were detected, where the maximum intensity is observed at the Z-value of

around -25, irrespective of the m/z values. Thus, a series of detected molecules in AM-VR may have a similar condensed aromatic-ring size with different lengths of alkyl side-chains. On the other hand, aromatic hydrocarbons, adducted with H^+ or Na^+, were detected as odd peaks by using ESI in methanol/chloroform/ trifluoroacetic acid (TFA) (acidic solution) in the range of $450 < m/z < 1000$.

2. Hydrocarbons and sulfur-containing aromatic hydrocarbons were detected in similar intensities by in-beam electron ionization (in-beam EI) in the range $100 < m/z < 650$. Hydrocarbons with low and high m/z regions are rich in aliphatic carbons and developed aromatic rings, respectively. The chemical formula analysis of all the sulfur-containing hydrocarbons detected in the range of $80 < m/z < 570$ indicates that the maximum appears at a similar Z-value of -10, which corresponding to benzothiophene derivatives.

3. Liquid secondary ionization (LSI) detects hydrocarbons with and without nitrogen- and sulfur-atom up to m/z 500. The most abundant constituents were one sulfur-containing compounds with carbon and Z-values of 17~18 and $-18~-22$, respectively.

Acknowledgements

This work has been carried out partially as a research project of The Japan Petroleum Institute commissioned by the Japan Cooperate Center, Petroleum with the subsidy of the Ministry of Economy, Trade and Industry.

References

1. Calemma, V.; Rausa, R.; D'Antona, P.; Montanari, L. *Energy & Fuels* **1998**, 12, 422-428.
2. Artok, L.; Su, Y.; Hirose, Y.; Hosokawa, M.; Murata, S.; Nomura, M. *Energy & Fuels* **1999**, 13, 287-296.
3. Wilt, B. K.; Welch, W.T. *Energy & Fuels* **1998**, 12, 1008-1012.
4. Strausz, O. P.; Mojelsky, T. W.; Lown, E. M. *Fuel*, **1992**, 71, 1355-1363.
5. Qian, K.; Tomczak, D. C.; Rakiewicz, F. F.; Harding, R. H.; Yaluris, G.; Cheng, W.-C.; Zhao, X.; Peters, A. W. *Energy & Fuels* **1997**, 11, 596-601.
6. Ueda, K.; Matsui, H.; Malhotora, R.; Nomura, M. *J. Jpn. Pet. Inst* **1991**, 34, 62-70.
7. Hsu, C. S.; Qian, K. *Energy & Fuels* **1993**, 7, 268-272.
8. Domin, M.; Herod, A.; Kandiyoti, R.; Larsen, J. W.; Lazaro, M-J.; Li, S.; Rahimi, P. *Energy & Fuels* **1999**, 13, 553-557.

9. Fujii, M; Yoneda, T.; Satou, M.; Sanada, Y. *J. Jpn. Pet. Inst.* **2000**, *43*, 149-156.
10. Marshall, A. G.; Henderickson, C. L.; Jackson, G. S. *Mass Spectrom. Rev.* **1998**, *17*, 1-35.
11. Hsu, C. S.; Liang, Z.; Campana, J. E. *Anal.Chem.* **1994**, *66*, 850-855.
12. Guan, S.; Marshall, A. G.; Scheppele, S. E. *Anal.Chem.* **1996**, *68*, 46-71.
13. Rodgers, R.P.; White, F. M.; Henderickson, C. L.; Marshall, A. G.; Andersen, K.V. *Anal.Chem.* **1998**, *70*, 4843-4750.
14. Qian, K.; Rodgers, R.P.; Hendrickson, C.L.; Emmett, M.R.; Marshall, A.G., *Energy & Fuels*, **2001**, *15*, 492-498.
15. Carr, S. R.; Cassady,C. J. *J. Mass Spectrometry*, **1997**, *32*, 959-967.
16. Senko, M. W.; Henderickson, C. L.; Emmett, M. R.; Shi, S. D. H.; Marshall, A. G. *J. Am. Soc. Mass Spectrom.* **1997**, *8*, 970-976.
17. Dell, A.; Williams, D. H.; Morris, H. R.; Smith, G. A.; Feeney, J.; Robert, G. C. K. *J. Am. Chem. Soc.*, **1976**, 97, 2497-2502.
18. Miyabayashi, K.; Suzuki, K.; Teranishi, T. ; Naito, Y.; Tsujimoto, K.; Miyake, M. *Chem. Lett.*, **2000**, 172-173.
19. Miyabayashi, K.; Naito, Y.; Tsujimoto, K.; and Miyake, M, *Eur. Mass Spectrom.*, **2000**, *6*, 251-258.
20. Fenn, J. B.; Mann, M.; Meng, C. K.; Wong, S. F.; Whitehouse, C. M. *Science*, **1989**, *246*, 64-71.
21. Qian, K.; Rodgers, R. P.; Hendrickson, C. L.; Emmett, M. R.; Marshall, A. G. *Energy Fuels*, **2001**, *15*, 492-498.
22. Miyabayashi, K.; Naito, Y.; Tsujimoto, K.; and Miyake, M. *Int.J. Mass Spectrum.*, **2002**, *221*, 93-105.
23. Bennlnghoven, A.; Sichtermann, W. K. *Anal. Chem.*, **1978**, *50*, 1180-1184.

Chapter 3

Thermal Properties and Dissolution Behavior of Ahphaltene and Resin

Y. Zhang[1], T. Takanohashi[1,*], S. Sato[1], I. Saito[1], and R. Tanaka[2]

[1]Institute for Energy Utilization, National Institute of Advanced Industrial Science and Technology, Tsukuba 305–8569, Japan
[2]Central Research Laboratories, Idemitsu Kosan Company Ltd., Chiba 299–0293, Japan

The thermal properties of asphaltene and resin were ascertained using differential scanning calorimetry (DSC) and thermogravimetry, and the heats of solution (ΔH_{solu}) of the two compounds in quinoline were established using microcalorimetry. The combination of the two techniques not only enables the nature of the interaction between asphaltene or resin and the solvent molecules to be evaluated but also provides a quantitative estimate as to its magnitude.

Introduction

Asphaltenes are very complex mixtures. They have a wide range of molecular weights and contain tens of thousands of compounds that have different functionalities. To date, characterization of asphaltenes has focused largely on chemical structural parameters, compositions, and aggregate sizes (*1-5*). However, an understanding of some of the processes that asphaltenes are involved in requires information as to their thermal properties in addition to knowledge of their chemical structure and composition. Thermal properties are particularly helpful in understanding the thermodynamic behavior of asphaltenes in various reactions.

Asphaltene molecules are known to form aggregates *via* various noncovalent interactions in crude oils or vacuum residues (VR), which is responsible for the formation of coke-precursors and the deactivation to catalytic reactions in upgrading and refining processes. In a recent study we suggested that coke formation could be reduced by dissociation of asphaltene aggregates in organic solvents (*6*). In order to control the relaxation of the aggregated structure in solvents effectively, it is important to fully understand the thermal properties of asphaltenes as well as the interactions that occur between asphaltene and solvent molecules.

In the present study the thermal properties of asphaltene (or resin) were determined using differential scanning calorimetry (DSC) and thermogravimetric analysis (TGA), along with an examination of their dissolution behavior in quinoline using microcalorimetry. By combining the two techniques a quantitative estimate of the interaction energy between the asphaltene and solvent molecules could be obtained.

Experimental

Samples and Chemicals

Asphaltene (AS) and resin (RE) were fractionated from Maya (MY) vacuum residues (VR) following vacuum distillation of the crude oil at 500°C. The properties of the samples are shown in Table I. Both asphaltene and resin occur as solid powders at room temperature.

Quinoline was purchased from Kanto Chem. Co., Inc., and used as received. *N,N*-dimethylnaphthylamine (DMNA) was supplied by Tokyo Kasei Kogyo Co., Ltd. The purity of this compound was >98% and its melting point was 41°C.

Table I. Properties of RE and AS

	Elemental analysis (wt%)					$\dfrac{H}{C}$	f_a	M_n
	C	H	S	N	O^a			
RE	82.4	9.5	5.7	1.1	1.3	1.38	0.39	770
AS	82.0	7.5	7.1	1.3	2.1	1.10	0.55	787

^{a}By difference. f_a: Aromaticity by ^{13}C-NMR.

Mn: Number average molecular weight by GPC.

Structural Characterizations

The aromaticity (f_a) and the number average molecular weight (M_n) of AS and RE were determined by ^{13}C-NMR and Gel permeation chromatography (GPC), respectively. ^{13}C-NMR spectra were measured in CDCl$_3$ with TMS as the internal standard on a Lambda 500 FT-NMR spectrometer at a proton resonance frequency of 600 MHz. Gel permeation chromatography (GPC) was carried out on an HPLC system equipped with a JASCO PU-986 pump, two polystyrene/polydivinylbenzene columns (Mixed-D type, 300 × 7.5 mm, Polymer Laboratories Ltd., U.K.), and a JASCO RI-930 differential refractometer detector using tetrahydrofurane (THF) as an eluent.

Differential Scanning Calorimetry (DSC) and Thermogravimetric Analysis (TGA)

The thermal properties of both samples were measured with a Seiko DSC 120 calorimeter. Temperature and enthalpy were calibrated with high purity indium and lead and the heat capacity was calibrated with synthetic sapphire. In a typical run, 6-10 mg of a sample was heated at a rate of 10°C/min from 8 to 300°C under nitrogen, introduced at a flow rate of 50 mL/min. After each initial scan, the sample was quickly quenched to 8°C and re-analyzed under the same conditions. Thermogravimetric analysis (TGA) was also carried out using a Shimadzu TGA-50H analyzer under the same experimental conditions as the DSC measurements.

Microcalorimetry Measurements

The heats of solution (or heat of mixing) of the samples in quinoline were measured using a micro-twin-calorimeter (MPC-11, Tokyo Riko Co., Ltd.). The detailed procedure of measurements and the examinations on the accuracy and reproducibility have been reported in our previous paper (7). A sealed glass ampoule containing the solid sample and a blank were prepared and immersed in sample and reference cells containing quinoline. The calorimeter was left overnight to reach thermal equilibrium, after which the measurement was started by simultaneously breaking the two ampoules within the two cells. In each dissolution measurement, the heat of solution was determined by mixing about 30 mg of solid sample with 20 ml of quinoline. The maximum operating temperature for this instrument is 60°C.

Results and Discussion

Theory and Methodology

Given that the aim of this study was to estimate the interaction energy between asphaltene and solvent molecules, any discussion of the results requires a theoretical framework. It is known that in the case of a liquid solute dissolving in a solvent, the heat of mixing (ΔH_{mix}) can directly reflect the essential interactions between solute and solvent molecules. If the mixing process is assumed to occur at a constant volume, then the heat of mixing is given by the Van Laar-Hildebrand equation (8) as:

$$\Delta H_{mix} = kT\chi_h \, n_1 \, \phi_2 \qquad (1)$$

where n_1 are the moles of solvent, ϕ_2 is the volume fraction of solute, and χ_h is the interaction parameter between the solute and solvent. However, when a solid solute dissolves in a solvent, the heat of solution (ΔH_{solu}) does not provide a direct interaction parameter insofar as it may be influenced by enthalpy changes arising from phase transitions, such as occur in the dissolution of glasses (9, 10), and/or changes in the crystalline structure (11). In this case, the heat of solution (ΔH_{solu}) can be expressed as the sum of the enthalpy changes due to any phase transition, as well as that of the heat of mixing (ΔH_{mix}). For example, in the case of completely crystalline materials,

$$\Delta H_{solu} = \lambda + \Delta H_{mix} \tag{2}$$

where λ represents the heat of fusion of the crystalline structure. While for ideal glass polymers,

$$\Delta H_{solu} = \Delta Cp \, (T - T_g) + \Delta H_{mix} \tag{3}$$

where T and T_g are an arbitrary temperature at solid state and glass transition temperature of the polymers, respectively, and ΔCp is the difference in heat capacity between the glass and solid states. The term $\Delta Cp \, (T - T_g)$ in Eq. 3 represents the glass enthalpy, which always yields an exothermic effect. In Eqs. 2 and 3, ΔH_{solu} can be determined directly by the microcalorimeter, while the heat of fusion and the glass enthalpy can be obtained from DSC measurements. Thus, it is expected that, for a solid material, the value of ΔH_{mix} may be estimated by a combination of microcalorimetry and DSC measurements.

Verification

To verify the reliability of the theory and methodology, *N,N*-dimethylnaphthylamine (DMNA) was used as a reference. This compound was selected because its melting point of 41°C is lower than the limit of the operating temperature of the microcalorimeter (60°C), thus allowing a direct comparison to be made between the determined and estimated values of ΔH_{mix}.

In the first instance, ΔH_{mix} for this compound was determined directly by using microcalorimetry at temperatures above its melting point. Figure 1 shows the microcalorimeter results for DMNA in

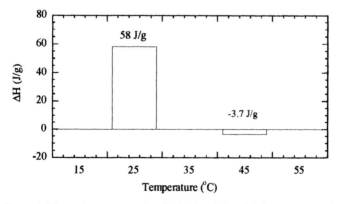

Figure 1. Microcalorimeter results of N,N-dimethylnaphthylamine in quinoline.

quinoline at different temperatures. The mixing of DMNA with quinoline at 45 °C proved exothermic responses with values for ΔH_{mix} of -3.7 J/g. This value is regarded as reflecting the magnitude of the interaction between DMNA and quinoline.

Secondly, ΔH_{mix} can be estimated from Eq. 2. As can be seen in Figure 1, when the microcalorimeter measurement was performed at 25 °C, i.e. below the melting point of DMNA, the heat of solution showed an endothermic effect with values for ΔH_{solu} of 58 J/g. It was confirmed that DMNA remained completely solid at 25°C. Consequently, the value of ΔH_{solu} at 25°C could be used to estimate ΔH_{mix}. Figure 2 shows the DSC thermogram of DMNA, which is typical of that obtained from a completely crystalline material. The heat of fusion (λ) was obtained by integrating the area of the endothermic peak on the DSC thermogram, giving a value of 61.4 J/g. From Eq. 2, ΔH_{mix} value is given as -3.4 J/g (exothermic) and is nearly the same as the analyzed values of -3.7. The similarity of the results demonstrates that a combination of DSC and microcalorimetry provides a useful tool for determining the ΔH_{mix} of solid materials, such as resin and asphaltenes, in solvents.

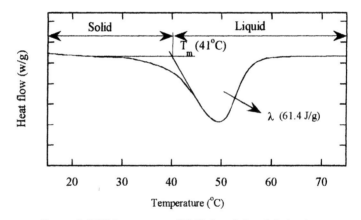

Figure 2. DSC thermogram of N, N-dimethylnaphthylamine.

Estimation of ΔH_{mix} of Resin with Quinoline

Compared with DMNA, resin (RE) is a very complex mixture of molecules that have different structures and molecular weights. Figure 3 shows a DSC thermogram of RE, which proved very similar to that of semi-crystalline materials in having both an endothermic peak and baseline shift (12, 13). This behavior suggests that RE contains both crystalline and glassy phases. In this case, the heat of solution should contain the contribution of both heats of fusion and glass enthalpy, i.e.,

$$\Delta H_{solu} = \lambda + \Delta Cp\ (T - T_g) + \Delta H_{mix} \qquad (4)$$

In this study we divided such DSC thermograms into two, as shown in Figure 3. The area α in Figure 3a corresponds to the heat of fusion of the crystalline phase. The glass enthalpy is calculated on the basis of a heat capacity (Cp) determination as shown in Figure 3b, which corresponds to the area surrounded by abe or estimated by multiplying ΔCp and ΔT (the area surrounded by $abcd$). The glass transition temperature (T_g) was defined at the mid-point of baseline shift on DSC or Cp curve, such as Figure 3, by the general methodology that has been previously described (14 – 16). The present results show that RE has a glass transition temperature of ca. 42°C i.e., sufficiently below the operating temperature limit of the microcalorimeter to allow direct determination of the heat of mixing at a temperature above the T_g of the resin.

Table II summarizes the microcalorimeter results for RE in quinoline at different temperatures, along with its thermal properties determined by DSC. These show that the ΔH_{mix} of RE in quinoline determined at 45°C is endothermic with a value of 4.3 J/g. The endothermic effect suggests that the resin molecules have little affinity with quinoline. On the other hand, from the thermal properties determined by DSC and from the heat of solution determined by microcalorimeter at 25°C (Table II), the ΔH_{mix} can be estimated as 3.3 J/g using Eq. 4. Both this estimated value and that determined directly by microcalorimeter (4.3 J/g) are sufficiently close as to suggest that estimations of the heat of mixing of complex materials such as resin may be made using a combination of DSC and microcalorimetric methods.

Estimation of ΔH_{mix} of Asphaltene with Quinoline

Asphaltene (AS) is heavier than resin. Elemental compositions and structural parameters are shown in Table I; they indicate that AS is more aromatic and contains more heteroatoms than RE. Although the two fractions show similar average values for their molecular weights, as determined by GPC, there is little doubt that AS has more condensed aromatic unit structures than RE, given that the former has a higher f_a value and a lower H/C ratio. These differences in structure and composition between asphaltene and resin might be expected to lead to different thermal responses and properties.

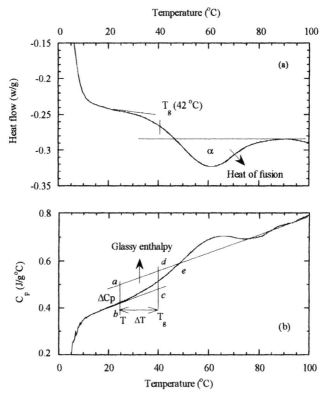

Figure 3. DSC thermogram (a) and heat capcity (b) of RE.

Table II. Thermal Properies, Heat of Solution, Determined and Estimated Heat of Mixing for RE and AS in Quinoline

	Tg^a	$\Delta C_p (T - T_g)^b$	λ^c	ΔH_{solu}^d	ΔH_{mix} Determined[e]	ΔH_{mix} Estimated[f]
RE	42	-1.8	4.6	6.1	4.3	3.3
AS	119	-11.8	5.6	-22.1	-	-15.9

[a]Glass transition temperature (^{o}C).   [b]Glassy enthalpy (J/g).   [c]Heat of fusion (J/g).
[d]Heat of solution measured at 25 ^{o}C (J/g).   [e]Heat of mixing measured at 45 ^{o}C (J/g).
[f]Heat of mixing estimated based on eq. 4.

Figure 4 shows the DSC thermogram of AS (*17*), which appears to indicate that the phase transition of AS begins at a temperature above 100°C. This means that for asphaltenes the ΔH_{mix} cannot be determined directly in its liquid state using microcalorimetry. With such materials, the combination of DSC and microcalorimetric methods that has been evaluated above in respect of DMNA and resin, offers an alternative way of determining ΔH_{mix}.

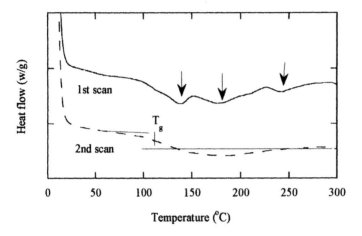

Figure 4. DSC thermograms of AS.

As shown in Figure 4, the DSC thermogram of AS is different from and far more complex than that of RE. Three endothermic events (indicated by arrows) occurred in the initial scan, but disappeared in the 2nd scan. These three events could have arisen from two patterns of heating behavior: the evaporation of volatile components or the release of adsorbed gases, and the softening of some components in the asphaltene. If evaporation contributes to the DSC scan, then a weight loss should be observed during heating.

Figure 5 shows the TGA and DTG curves of AS. In the initial heating run, AS began to lose weight near 100°C. A broad inflection was recorded from 100 to 200°C on the corresponding DTG curve i.e., over a similar temperature range to the first and second endothermic events observed in the initial DSC scan (Figure 4) which, presumably, reflects the observed weight loss resulting from these endothermic events. Consequently, both the evaporation of light components and the melting of crystallized phase in asphaltenes contribute to the events recorded in the initial DSC scan over the temperature range analyzed. Such an overlapping of effects makes it difficult to glean accurate information about the thermal properties of asphaltenes based on this initial DSC scan alone. However, the 2nd scan shows that an obvious baseline shift occurred between temperatures below 100°C and those above 230°C (Figure 4), which provides clear evidence that asphaltenes undergo a glass transition. At the same time, another broad endothermic event, centered around 180°C, indicates that AS also contains crystalline phase(s). Thus, because of the weight loss in the first DSC scan, the thermal properties of AS were determined from the second DSC run and used to estimate the heat capacity. The results are summarized in Table II.

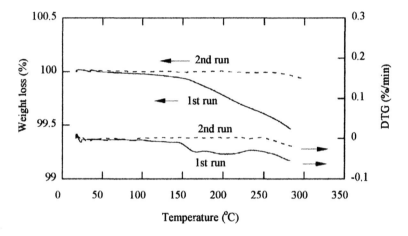

Figure 5. TGA and DTG curves of AS.

Table II also shows the heat of solution for AS in quinoline at 25°C. From the thermal properties determined by DSC and from the heat of solution in quinoline at 25°C, the value of ΔH_{mix} was estimated at -15.9 J/g based on Eq. (4) for AS. The exothermic nature of ΔH_{mix} suggests that AS is strongly solvophilic with respect to quinoline.

Conclusions

The experimental results obtained from the differential scanning calorimetry and microcalorimetry measurements of the present study yield the following conclusions. Asphaltene and resin contain both glassy and crystalline phases. Asphaltene exhibited exothermic behavior and proved solvophilic to quinoline. Resin showed endothermic activity and was solvophobic to quinoline. The combination of differential scanning calorimetry and microcalorimetry provides a quantitative measure of the interaction between asphaltenes and organic solvents. For example, the heat of mixing of asphaltene in quinoline was determined successfully as −15.9 J/g from ascertaining the heat of solution and measuring thermal properties by DSC.

References

1. Su, Y.; Artok, L.; Murata, S.; Nomura, M. *Energy Fuels*, **1998**,*12*, 1265.
2. Sheu, E. Y.; Ling, K. S.; Sinha, S. K.; Overfield, R. E. *J. Colloid Interface Sci.* **1992**, *153, 399*.
3. Miller, J. T.; Fisher, R. B.; Thiyagarajan, P.; Winans, R. E.; Hunt, J. E. *Energy Fuels*, **1998**, *12*, 1290.
4. Long, R. B.; Speight, J. G. *Rev. Inst. Fr. Pet.* **1990**, 45, 553.
5. Wiehe, I. A. *Ind. Eng. Chem. Res.* **1992**, 31, 530.
6. Takanohashi, T.; Sato, S.; Tanaka, R. *Petr. Sic. Technol.*, **2003**, *21*, 491.
7. Zhang, Y.; Takanohashi, T.; Sato, S.; Saito, I. Tanaka, R. *Energy Fuels*, **2003**, *17*, 101.
8. Hidebrand, J. H., Scott, R. L. *The Solubility of Non-Electrolytes*, Reinhold, New York, **1950**, Chapter 20.
9. Bianchi, U.; Pedemonte, E.; Rossi, C. *Markromol. Chem.*, **1966**, *92*, 114.
10. Cottam, B. J.; Cowie, J. M. G.; Bywater, S. *Makromol Chem.*, **1965**, *86*, 116.
11. Schreiber, H. P.; Waldman, M. H. *J. Polymer Sci.* **1967**, *A2, 5*, 555.
12. Wunderlich, B.; Boller, A.; Okazaki, I.; Ishikiriyama, K.; Chen, W.; Pyda, M.; Park, J.; Moon, I; Androsch, R. *Thermochim. Acta*, **1999**, *330*, 21.
13. Wunderlich, B.; Okazaki, I.; Ishikiriyama, K.; Boller, A. *Thermochim. Acta*, **1998**, *324*, 77.
14. Claudy, P.; Létoffé, J-M.; Chagué, B.; Orrit, J. *Fuel*, **1988**, *67*, 58.
15. Hansen, A. B.; Larsen E.; Pedersen, W. B.; Nielsen A. B.; RØenningsen, H. P. *Energy Fuels*, **1991**, *5*, 914.
16. Claudy, P.; Létoffé, J-M.; King, G. N.; Planche, J. P.; Brûlé, B. *Fuel Sci. Technol. Int.* **1991**, *9*, 71.
17. Zhang, Y.; Takanohashi, T.; Sato, S.; Saito, I. Tanaka, R. *Energy Fuels*, **2004**, *18*, 283.

Chapter 4

Characterization of Refractory Sulfur Compounds in Petroleum Residue

I. Guibard, I. Merdrignac, and S. Kressmann

[1]Instut Français du Pétrole, IFP-Lyon, BP3, 69360 Vernaison, France

Sulfur removal is an important step in the processing of petroleum residue into more valuable products and fuel oil production. The demand for low sulfur fuel oil (from 0.5 to 0.3 weight %) requires improving knowledge of sulfur compounds in heavy oils in order to find the best association of hydrotreating catalysts. However, the high boiling point of residues makes it difficult to characterize and to identify the chemical structure of the most refractory sulfur species. The use of conventional analyses like liquid chromatography separation and elemental analysis gives a first approach of the location of refractory compounds. Moreover, the Size Exclusion Chromatography (SEC) enables to follow the evolution of the relative size distribution of the feed and effluent during hydroconversion. In this work, several feeds and effluents were characterized and information on the evolution of heavy sulfur compounds was obtained. This approach helps to select the best catalyst combination taking into account the feed composition and its reactivity.

51

52

The refining of petroleum residues is more and more focussed on the removal of sulfur compounds contained in fuel oils. This low sulfur fuel oil (LSFO) is essentially used in power plants for electricity production that enables to reduce SOx emissions. Moreover desulfurized fuel oil is necessary to have a suitable feed for Resid Catalytic Cracking (RCC) units in order to increase gasoline yield and to reduce SOx emissions and catalyst consumption. Atmospheric residues (AR) and above all vacuum residues (VR) are the most difficult feeds to process catalytically as they contain and concentrate most of the crude impurities. Amongst these are asphaltenes, metals and heavy sulfur compounds, which affect the process performances.

These residues are generally hydrotreated in a fixed bed unit using high-pressure hydrogen and specific catalysts. The process consists of two complementary sections installed in series: the demetallization (HDM) section and the refining section (HDS). The main objective of the HDM section is to disaggregate asphaltenes and to remove most of the metals. During this step a partial hydrodesulfurization (HDS) also occurs. In the HDS section the deeply transformed feed is refined to remove most of the sulfur content from the effluent and to reduce Conradson Carbon in order to produce a relevant RCC feed. Depending on the case, an intermediate section (reactor 3 in Figure1) could be included with a specific catalyst, which is able to continue the demetallization and to begin the desulfurization step (1,2). For each section, specific catalysts were developed in order to achieve high severity levels and high duration cycle lengths. The IFP Hyvahl residue upgrading process developed in 1982 (Figure 1) has a well adapted graded catalyst system that is able to cope with a wide range of feeds and to associate high severity levels together with constant product quality and increased cycle lengths (2).

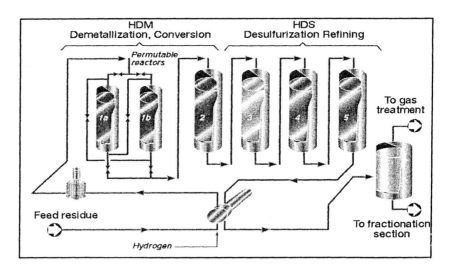

Figure1. IFP Hyvahl process

For a few years now, the demand for very low sulfur fuel oil (from 0.5 to 0.3 weight %) during long cycle has increased on the worldwide market. This objective is difficult to achieve and requires increasing knowledge of heavy refractory sulfur compounds in various feeds and kinetic information about the sulfur compound reactivity. But the characterization of sulfur compounds in fuel oil is complex due to the high boiling range of these products and the unavailability of a detailed analytical technique as sulfur speciation for diesel.

Experimental Section

Feeds and Hydrotreatment Experiments

A first set of experiments was performed on Middle East residues in order to locate sulfur compounds. Several feedstocks (AR, VR) were hydrotreated in a fixed bed reactor unit under the same standard operating conditions to achieve the target of 0.3 wt% sulfur content in fuel oil. Experiments were conducted using industrial residue hydrotreatment catalysts (HMC841 and HT series from Axens). The characteristics of these feedstocks are given in Table I

Table I Main characteristics of Middle East feedstocks

	Arabian Medium AR	Arabian Light VR	Arabian Heavy VR
Type	II	II	II
d15/4	0.981	1.029	1.048
Total S (wt%)	3.7	4.5	5.3
Total Ni+V (ppm)	80	125	218
Asph (wt%)	6.3	11.5	16.2
S (wt%)	7.2	6.8	7.6
Res 520°C+ (wt%)	20.0	34.2	34.3
S (wt%)	5.7	5.4	6.4
Aro 520°C+ (wt%)	20.0	43.7	40.7
S (wt%)	3.8	4.0	4.5
Sat 520°C+ (wt%)	4.3	7.8	4.2
S (wt%)	0.1	0.9	-
520°C- cut (wt%)	49.4	2.8	4.6
S (wt%)	2.6	3.5	3.9

Note: -Average value for asphaltene content

A second set of experiments was carried out only on HDM catalyst with varying residence time in order to study the HDS performance precisely. Other residues of different types of organic matter were used like Duri VR (type I: lacustrine origin), Ural VR (type II: marine origin) and Ardjuna VR (type III: terrestrial origin). The characteristics of these feedstocks are given in Table II

Table II. Main characteristics of others feedstocks

	Duri VR	*Ardjuna VR*	*Ural VR*
Type	I	III	II
d15/4	0.963	0.987	1.005
Total S (wt%)	0.6	0.3	2.7
Total Ni+V (ppm)	75	27	220
Asph (wt%)	5.7	3.5	5.3
S (wt%)	*1.24*	*0.78*	*3.27*
Res 520°C+ (wt%)	44.7	41.8	33.4
S (wt%)	*0.87*	*0.43*	*2.79*
Aro 520°C+ (wt%)	28.6	36.1	43.1
S (wt%)	*0.56*	*0.23*	*3.92*
Sat 520°C+ (wt%)	21.0	18.6	13.5
S (wt%)	-	-	-
520°C- cut (wt%)	-		4.7
S (wt%)			*2.19*

Note: -Average value for asphaltene content

Analytical Techniques

Liquid products were distillated in three cuts (PI-375, 375-520°C and 520°C+). The heaviest cut is separated into asphaltene (Asph) and maltene fractions using n-heptane. The maltene fraction and the 375-520°C cut were further fractionated with a liquid chromatography column into saturates (Sat), aromatics (Aro) and resins (Res). On each SARA fraction, elemental analysis (Carbon, Hydrogen, Nitrogen, Sulfur, Oxygen) was performed. Metal (Nickel and Vanadium) concentrations in asphaltenes were measured by ICP method (Figure 2).

Size exclusion chromatography (SEC) was performed on the different fractions on a Waters 150CV+ using a refractive index detector. Calibration was carried out using polystyrene standards of masses from 162 to 120000 g/mol. Samples are injected in THF (5g/l) at 40°C and at a flow rate of 1ml/min. The column packing was PS-DVB (polystyrene-divinylbenzene). Each operating parameter was fixed in order to obtain a precise and reproductible SEC molecular weight distribution. However, the SEC technique is only able to give a relative result (3).

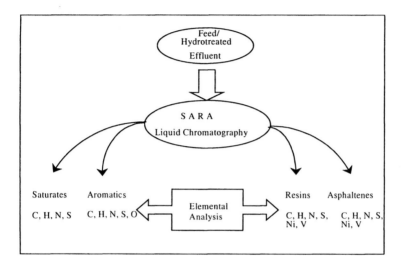

Figure 2 Analytical protocol

Sulfur Evolution during Hydrotreatment

In atmospheric or vacuum residue, sulfur is initially concentrated in the heaviest fraction (cut 520°C+) and specifically in asphaltenes, which are described as the most refractory class to hydrodesulfurization *(4,5)*. During hydrotreatment, the total sulfur content decreases all along reactors and can reach 0.3 %wt at the outlet of the unit under appropriate operating conditions (temperature, residence time). At the same time, residue conversion occurs in HDM section and heavy fractions as asphaltenes and resins decrease to produce lighter fractions (Figure 3).

In HDM section, aromatic fraction remains relatively constant. In fact some heavy aromatic compounds are converted into lighter fractions, but other aromatics are produced from asphaltenes and resins hydrocracking. In HDS section, the use of specific catalyst leads to a slight decrease of heavy aromatic fraction and asphaltenes are not removed due to the mesoporous structure of the

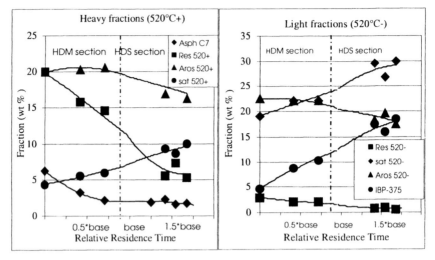

Figure 3. Arabian Medium AR Hydrotreatment.
Fraction (SARA) evolution in heavy cut (520°C+) and light cut (520°C-)

HDS catalyst inducing intragranular diffusion phenomena. For resins, hydrocracking continues to occur but seems to slow down even at high residence time. For lighter fractions (520°C- cut), a continuous elimination of resins is observed when light aromatics remain constant in HDM section and decrease more distinctly in HDS section.

Figure 4 shows the evolution of the sulfur content inside the SARA fractions versus residence time for the HDM and HDS section, in the case of Arabian Medium AR hydrotreatment. All fractions are continuously desulfurized along reactors but at different rates. Thus for HDM section, we observe an important decrease of sulfur content for all the fractions. The HDM catalyst is able not only to disaggregate the asphaltenes but also to eliminate a great part of sulfur compounds included in this fraction.

In the HDS section, the desulfurization rate seems lower for each fraction but the removal of ultimate sulfur is much more difficult as it is often described in literature *(6,7)*. The sulfur contents in resin and aromatic fractions reach a low level due to the use of a specific catalyst whereas it still remains at a high level in the asphaltenes. This indicates that the sulfur is located in high molecular weight and does not access to the active sites in the porosity of the HDS catalyst. However, this specific porosity is necessary for efficient removal of sulfur in lower molecular weight fractions.

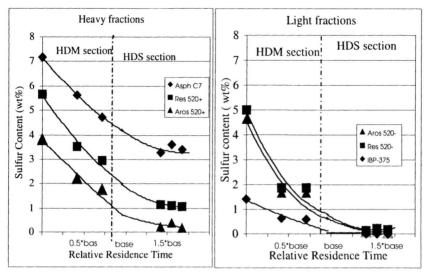

Figure 4. Arabian Medium AR hydrotreatment
Sulfur content evolution in heavy cut (520°C+) and light cut (520°C-)

After HDM and HDS section, the residual sulfur (0.3 wt%) is still concentrated in the 520°C+ fraction. The sulfur distribution is about 15% in asphaltenes, 40% in resin and aromatic fractions, 5% in light compounds (520°C-). Therefore the sulfur is essentially concentrated in heavy aromatic and resin fractions. An improvement of the catalyst layout is the use of a catalyst with balanced HDS/HDM activity after HDM catalyst in order to have a more efficient sulfur removal in heavy resins and aromatics.

The sulfur compounds in the asphaltene fraction cannot be eliminated even at a high residence time in the HDS section. Consequently the desulfurization of asphaltenes is essentially done in the HDM section. The same observation has been made for VR hydrotreatment.

In order to better understand asphaltene behavior, we have studied the evolution of sulfur compounds in the HDM section for different relative residence times. Experiments are carried out on Arabian Heavy VR for the HDM section. We can observe an apparent slowing down of resins and aromatics desulfurization (Figure 5) because the HDM catalyst has not a sufficient hydrogenation function and a high surface area unlike the HDS catalyst. The same experiments were conducted with deasphalted oil (DAO/C7) obtained from n-heptane precipitation of the same vacuum residue. The decrease of sulfur content in resin and aromatic fractions after DAO/C7 hydrotreatment is more important than for vacuum residue (Figure 5). This shows the inhibitor effect of asphaltenes for hydrodesulfurization reaction which, could be due to a strong adsorption of these heavy molecules on the catalyst surface that block the access to the active sites for resins and aromatics. However aromatics and resins produced by asphaltene hydrocracking may be also more refractory than the initial fractions, this could explain the difference between DAO and VR reactivity and desulfurization rates (8,9).

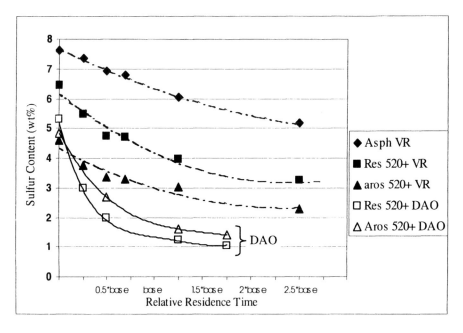

Figure 5. Arabian Heavy VR and DAO/C7 hydrotreatment in HDM section. Evolution of sulfur content in 520°C+ fractions

In Figure 5, we can also observe, even for DAO/C7 feed a slow down of HDS reactions at high relative residence times. This can be due to the presence of refractory compounds, even in this deasphalted feed. Thus, the introduction of another catalyst appears necessary. The proportion of each catalyst (HDM and HDS catalysts) needs to be optimized according to the nature of the feedstock and the performances required. Then, this curve of the sulfur content evolution is very useful for this optimization.

In conclusion, asphaltenes appear like the most difficult class to desulfurize and also have a strong inhibitor effect on desulfurization reactions due to the production of refractory compounds by asphaltenes disaggregation. Moreover, the presence of asphaltenes on the catalytic surface probably slows down the desulfurization rate of resin and aromatic fractions.

SEC Analyses

In order to obtain more information about evolution of heavy compounds, Size Exclusion Chromatography (SEC) has been performed on asphaltene fraction in Arabian Medium AR before and after hydrotreatment. Profiles of relative distribution size are reported in Figure 6. Bimodal distributions are observed probably because of the asphaltenes repartition between associated (high molecular weight) and less associated species (low molecular weight). However, the low molecular weight tailing of the chromatogram may come from non size-effects, which basically are due to adsorption of species on the stationary phase of the column.

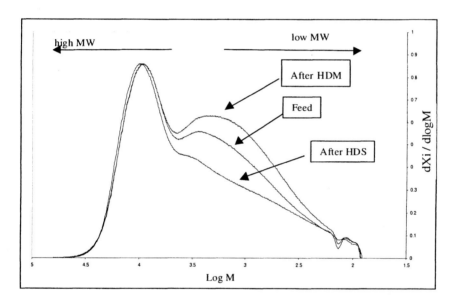

Figure 6. Arabian Medium AR hydrotreatment.
SEC chromatograms of non converted and converted asphaltenes after HDM and HDS section in the Hyvahl process

Results show that the structure of high molecular weight molecules is not affected in the HDM section, but an increase of the low to high molecular weight peaks ratio is observed due to a conversion of high molecular weight compounds into smaller species and a dealkylation of side alkyl chains. During this step either smaller molecules can be formed or structural modifications may occur. After the second step of the process (HDS), chromatogram profiles show however a significant decrease of the low molecular weight peak. Interpretations of such results may be a preferential conversion of small asphaltene molecules into resins and aromatics due to the high level of hydrogenation of HDS catalyst. Besides, high molecular weight population seems to be equivalent after the first and the second step of the process. In conclusion, the highest molecular weight asphaltenes that are not eliminated on HDM catalyst are still remaining at the outlet process after HDS. This confirms the kinetic evolution of sulfur in asphaltenes.

Influence of the Crude Oil Origin

The nature and type of feedstocks are very important parameters in hydrotreatment processes regarding catalyst performances. Asphaltenes content, metal impureties level and size of asphaltenes are generally critical properties for catalyst activity. In Tables III and IV, characteristics of different residues are reported and also elemental analyses of asphaltene fraction in feed and in hydrotreated effluent (at the HDM section outlet).

Figure 7 describes the evolution sulfur removal in asphaltene fraction after HDM section versus the initial asphaltenes content of the feed. Thus, for fixed and identical operating conditions, a curve allows to describe Middle East feed behavior. So, asphaltenes content appears to be the most important parameter to evaluate feed reactivity for HDS (1).

Table III. Performances in HDM section for Middle East feedstocks

	Arabian Light AR	Arabian Light VR	Arabian Heavy AR	Arabian HeavyVR
Type	II	II	II	II
d15/4	0.971	1.029	0.985	1.048
S (wt%)	3.4	4.5	4.6	5.3
Ni+V (ppm)	53	154	165	218
Asph (wt%)	3.5	11.5	9.1	16.8
S (wt%)	6.82	6.76	7.47	7.6
N (wt %)	0.885	0.955	1.07	0.85
H/C	1.05	1.05	1.09	1.07
HDM section outlet				
Asph (wt%)	1.8	5.7	5.3	9.7
S (wt%)	4.91	5.97	6.42	6.93

Notes: -Performances on HDM stabilized catalyst for identical operating conditions
-Average value for asphaltene content

Table IV. Performances in HDM section for different residues

	Duri VR	Ardjuna VR	Ural VR
Type	I	III	II
d15/4	0.963	0.987	1.005
S (wt%)	0.6	0.28	2.7
N (ppm)	5602	3098	5800
Ni+V (ppm)	75	28	220
Asph (wt%)	5.7	3.5	5.3
S (wt%)	*1.24*	*0.78*	*3.27*
N (wt%)	*1.54*	*0.97*	*1.42*
H/C	*1.1*	*1.04*	*1.04*
HDM section outlet			
Asph (wt%)	3.6	1.6	3.9
S (wt%)	*1.14*	*0.83*	*2.73*

Note: Performances on HDM stabilized catalyst for identical operating condition

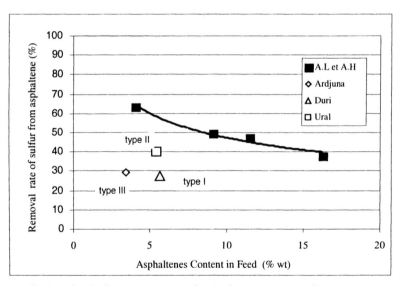

Figure 7 : Residue hydrotreatment at identical operating condition
Removal rate of sulfur from asphaltene fraction versus total asphaltenes content

However for others types of organic matter (type I and type III), and also for an other origin of type II crude (Ural VR) this relation seems different. For these residues, the desulfurization of asphaltenes is really lower than for Middle East feeds, which can be due to either a specific sulfur compounds, or to the asphaltenes size.

In order to study specifically these differences, profiles of relative distribution asphaltenes size were obtained by SEC for these different residues (Figures 8 and 9). The mass distributions of these asphaltenes are bimodal again but present different profiles. Most of the asphaltenes samples are dominated by high molecular masses and/or associated structures, except in the case of Ardjuna where high and low molecular weight peaks are equivalent. Asphaltenes from Duri show a distribution of higher mass molecules due probably to the existence of long alkyl chains (type I) illustrated by a high H/C ratio (asphaltene H/C=1.1). On the contrary, asphaltenes from Ardjuna (type III) do not present any long paraffinic chains (H/C=1.04) and consequently the SEC profile is shifted towards lower masses. The low reactivity of Duri could be explained by a different size distribution and by the quality of organic matter. For Ardjuna, despite of lower asphaltenes size, the HDS activity is low. It may be due to the organic matter, which is more aromatic (mainly lignite stuctures (type III)).

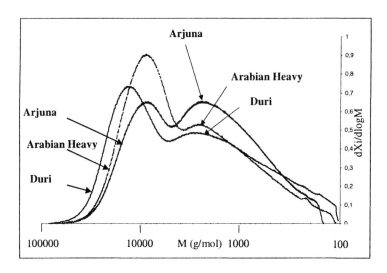

Figure 8 SEC Chromatograms of VR Asphaltenes from different sources

SEC profiles of various Middle East asphaltenes are relatively similar. This explains an equivalent behavior in hydrotreatment (Figure 9). For Ural, asphaltenes profile is not really different from Middle East whereas the reactivity of sulfur compounds is lower. This could be explained by the high concentration of nitrogen in Ural asphaltenes, which can have an inhibitor effect on the HDM catalyst *(6)*.

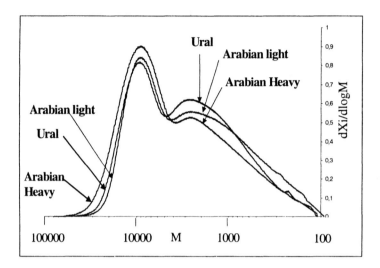

Figure 9.SEC Chromatograms of VR Asphaltenes from various type II crude oil

Finally, profiles of distribution asphaltene size allow to differentiate residue and their behavior in hydrodesulfurization. But, an equivalent profile is not sufficient to conclude that the HDS reactivity is similar. Then, other parameters, such as sulfur compounds structure, asphaltenes structures and origin of crude oil should be considered.

Conclusion

The increasing severity of sulfur specifications for fuel oil calls for the need of more knowledge about the heaviest sulfur compounds. Deep desulfurization can be achieved with appropriate catalytic system and operating conditions. Usual analyses like liquid chromatography separation and elemental analysis give a first approach of the location of refractory compounds and can be a help to better understand the specific role of each catalyst for HDS reactions. So, this study demonstrates that after a deep hydrotreatment, the sulfur is essentially concentrated in heavy aromatic and resin fractions (cuts 520°C+). The desulfurization of asphaltenes is difficult and occurs mainly in the HDM section. Moreover the desulfurization rates of other fractions slow down because of the presence of the asphaltenes on the catalytic surface and the production of more refractory compounds by asphaltenes disaggregation.

The SEC method used enables to differentiate asphaltenes from different sources and to follow feed and effluent conversions. It is a useful technique in the understanding of hydroprocessing process, but gives only relative molecular weight distributions. However, this analysis is not sufficient to conclude on the HDS reactivity. The origin of the organic matter and probably the nitrogen content of the crude must be considered.

The development of new appropriate analytical methods seem to be required to isolate and characterize heavy sulfur compounds particularly in resin and aromatic fractions. This analysis should give a better understanding of kinetic HDS reactions and should be a real help to define the best catalyst layout depending on the feed composition.

Acknowledgements: We would like to thank F.X. Haulle for his contribution during his Ph.D at IFP.

References

1   Kressmann, S.; Morel, F.; Harlé, V.; Kasztelan, S. *Catal Today* **1998**, 43, 203-215

2.   Kressmann, S.; Harlé, V.; Kasztelan, S.; Guibard, I.; Tromeur, P.; Morel, F. Prepr. Pap - Am Chem Soc., Div Fuel Chem, 1999

3.   Merdrignac, I.; Truchy, C.; Robert, E.; Desmazieres, B.; Guibard, I.; Haulle, F.X.; Kressmann, S. Heavy Organics Deposition Conference (HOD), Puerta Vallarta, 2002

4.   Speight, J.G. In *The Desulfurization of heavy oils and residua 2nd edition*; Dekker, M.,Inc : New York, 2000; pp. 127-167.

5.   Bartholdy, J.; Andersen, S.I *Energy Fuels* **2000**, 14, 52-55

6.   Callejas, M.A.; Martinez, M.T. *Energy Fuels* **2000**, 14, 1304-

7.   Seki, H.; Kumata, F. *Energy Fuels* **2000**, 14, 980-85

8.   Marafi, A.; Stainlaus, Q.; Hauser, A.; Fukase, S.; Matsushita, K.; Al-Barood, A.; Absi-Halabi, M. Prepr. Pap - Am Chem Soc., Div Fuel Chem, 2001

9.   Haulle, F.X.; Kressmann, S. Aiche Spring Meeting, New Orleans, 2002

Chapter 5

Estimation of the Structural Parameter Distribution of Asphaltene Using a Preparative GPC Technique

Shinya Sato[1], Toshimasa Takanohashi[1], and Ryuzo Tanaka[2]

[1]Institute for Energy Utilization, National Institute of Advanced Industrial Science and Technology, 16–1 Onogawa, Tsukuba, Ibaraki 305–8569, Japan
[2]Central Research Laboratory, Idemitsu Kosan Company Ltd., 1280 Kami-izumi, Sodegaura, Chiba 299–0393, Japan

Vacuum residues of three kinds of asphaltenes, Iranian Light, Khafji, and Maya, were separated into 5 fractions by a preparative GPC, followed by an average structural analysis to estimate the structural parameter distribution. The H/C atomic ratio increased and carbon aromaticity (fa) decreased as the molecular weight (MW) increased up to MW of 1700 and was nearly constant at MW above 1700, while H/C + fa was constant throughout the range of MW. The parameters on the numbers of total and aromatic rings, N/C and S/C were nearly constant. The estimated formula weight of a unit was between 400 and 1200. The results suggest that asphaltene molecules can be classified into 3 types according to MW: (1) low MW molecules, consisting of a single unit, (2) moderate MW molecules, consisting of 1-2 units, and (3) high MW molecules, consisting of several units with no further enlargement of a unit size, which suggests polymerization or aggregation of the units.

The structure of asphaltene is an important factor for investigating the reactivity of heavy hydrocarbons. Although asphaltene displays a variety of properties due to a complex mixture of molecules with a wide range of molecular weights (MW), most conventional discussions are based on the average molecular structure. Some researchers have tried to reflect structural distributions into structural parameters, but most are expressions of parameters such as molecular weight with an assumption of a mathematical distribution function[1] and not an expression of distribution based on the analytical results. It is believed that asphaltene has high aromaticity when the average MW is low and a low aromaticity when the average MW is high[2]. Thus, the properties of asphaltene are thought to be dependent on MW, but this tendency is unclear.

In the present study, asphaltene was fractionated according to the molecular weight using gel permeation chromatography (GPC). Then average molecular structural analyses were conducted on each fraction in order to clarify the dependency of the structural parameters on the MW.

Experiment

This study used asphaltene feeds recovered from the vacuum residues of Iranian Light crude oil (AS-IL) from Iran, Khafji crude oil (AS-KF) from Saudi Arabia, and Maya crude oil (AS-MY) from Mexico, whose properties are shown in Table I.

The apparatus used for the fractionation was a GPC system (PU-980 HPLC pump, JASCO Co Ltd.) equipped with an auto injector (AS-950-10), an ultraviolet detector (UV-970), a fraction collector (SF-212N), and a preparative column (Shodex KF-2003, 20mm of inner diameter and 300mm length, exclusive limit of 70,000). For the fractionation, 2ml of a solution containing 1 wt% of asphaltene in chloroform was charged into an injection loop, then fractionated into 30 fractions, which covered a MW range between 70,000 and 100, using 3ml/min of chloroform as an eluent. The injection was repeated every hour, while the eluent was fractionated into 30 fractions from 12 to 27 min. every 0.5 min. by retention time. The fractionation for each feed was repeated 50 times to separate 500mg of asphaltene. After the fractionation, the amount of asphaltene in each fraction was determined by an analytical GPC system equipped two columns linked in series (Shodex KF-403HQ, 4mm of inner diameter and 300mm length) and an evaporative laser scattering detector (Varex MK III ELSD, Alltech Associates Inc.). Then the collected fractions were consolidated into 5 fractions (Frs. 1 to 5) in order to recover enough samples for ^{13}C-NMR analysis. In addition, uncollected samples were recovered from waste eluent (Fr. 6), which was eluted after 30 fractions. Fr. 6, which recovered from about 8 L of eluent, contained a certain amount of phthalates probably dissolved

Table I Properties of Feedstocks and Fractionated Asphaltenes

Fraction	Feed	1	2	3	4	5	6[*1]
Asphaltene from Iranian Light vacuum residue (AS-IL)							
Recovery, wt%	-	17	17	17	16	15	18
MW[*2](GPC)	841	5682	3029	1608	658	189	4735
Elemental analysis							
C	83.2	80.8	82.8	82.9	83.2	82.7	79.5
H	6.8	7.8	7.8	7.8	7.5	6.8	7.1
N	1.4	1.5	1.5	1.5	1.6	1.7	1.4
S	5.9	5.3	5.4	5.3	5.3	5.4	7.1
H/C atomic ratio	0.98	1.14	1.12	1.12	1.07	0.98	1.06
fa[*3]	0.57	0.48	0.48	0.48	0.53	0.60	0.55
Ashaltene from Khafji vacuum residue (AS-KF)							
Recovery, wt%	-	13	14	11	10	13	39
MW[*2](GPC)	944	7856	3617	1712	725	221	2230
Elemental analysis							
C	82.2	79.9	81.3	81.3	82.2	80.4	75.5
H	7.6	7.5	7.8	7.7	7.4	6.8	7.2
N	0.9	1.0	1.0	1.0	1.0	1.1	1.1
S	7.6	7.6	7.6	7.4	7.4	7.8	6.9
H/C atomic ratio	1.10	1.12	1.13	1.13	1.08	1.01	1.14
fa[*3]	0.50	0.48	0.48	0.47	0.51	0.56	0.53
Asphaltene from Maya vacuum residue (AS-MY)							
Recovery, wt%	-	17	13	13	16	10	31
MW[*2](GPC)	943	6984	3318	1677	674	145	3165
Elemental analysis							
C	82.0	79.4	81.5	81.7	81.6	81.6	80.5
H	7.5	7.9	7.9	8.1	7.7	7.0	7.1
N	1.3	1.2	1.2	1.4	1.6	1.6	1.5
S	7.1	7.2	7.5	6.9	5.3	6.7	5.6
H/C atomic ratio	1.09	1.18	1.16	1.18	1.12	1.02	1.05
fa[*3]	0.53	0.46	0.49	0.49	0.53	0.56	0.50

*1 Fraction recovered from waste eluent
*2 Number averaged molecular weight
*3 Carbon aromaticity

from glassware. Phthalates were removed by washing with heptane until no peak was observed on ^{1}H-NMR spectrum

Subsequently, in order to investigate the molecular weight dependency of the representative parameters of the average structure, average structure analyses[3] were conducted from the results of the MW measurements using a GPC, elemental analysis, and ^{1}H- and ^{13}C-NMR analysis[4].

Results and Discussion

GPC separation

The recovery yields of the fractionated asphaltene summarized in Table I were calculated based on the total recovery amounts of Frs. 1-5 and Fr. 6. The recoveries for Fr. 6 were 18wt% for AS-IL, 39wt% for AS-KF and 31 wt% for AS-MY. The existence of Fr.6 suggests that there are any other mechanisms on GPC separation besides only molecular size. It is probably corelated to fa. Further analysis may be necessary for the mechanism.

Properties of asphaltene

Table I also summarizes the properties of feeds and each fraction. Here, carbon content of feed is higher than the average of the fractions. One of the considerable reason is that a small part of chloroform insolubles with higher carbon content were eliminated by prefiltration for GPC, or that remained in a GPC column eternally.

The number averaged MW determined by GPC was distributed between 100 and 8000. However, it is believed that MW by GPC does not represent an accurate MW since a polystyrene standard was used for the calibration curve between retention time and MW. In order to eliminate the error caused by the inaccuracy of MW, the structural parameters per carbon are discussed hereafter.

The distributions of H/C atomic ratio and carbon aromaticity (fa) to MW by GPC suggests that there is a point of inflection around MW of 1700 (Figure 1). Up to a MW of 1700, the H/C atomic ratio increased as the MW increased (Frs. 3-5), but remained constant above 1700 (Frs. 1-3). The tendency was common for all asphaltenes, although that of AS-MY was slightly larger than the others. N/C and S/C remained constant throughout the MW range. The carbon aromaticity (fa) showed a complementary distribution to the H/C atomic ratio, i.e., fa decreased as the MW increased up to approximately1700 and remained constant above 1700.

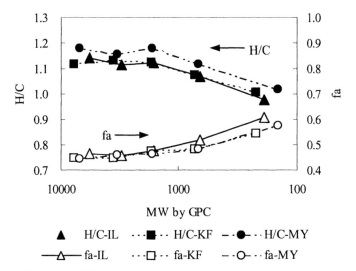

*Figure 1 H/C atomic ratio and carbon aromaticity (fa) for
fractionated asphaltenes*

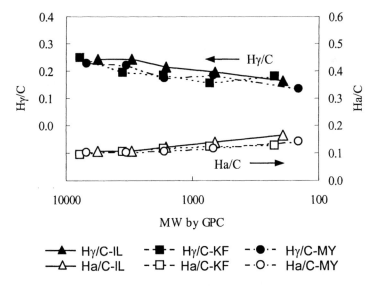

*Figure. 2 Patameters of aromatic hydrogen (Ha/C) and terminal
methyl group (Hγ/C) for fractionated asphaltenes*

The distributions of hydrogen aromaticity (Ha/C) slightly increased as the MW decreased, but the variation was about 0.05, less than H/C and fa (Figure 2). and hydrogen on terminal methyl per carbon (Hγ/C) decreased about 0.1 as the MW decreased from 10,000 to 100. Both showed linear correlations to the logarithm of MW.

Fr. 6, which had a MW similar to Fr. 2, did not fit those correlations for every asphaltene as it had a lower H/C and a higher fa except for AS-KF, which suggests molecules with more aromaticity in Fr. 6. This may be one of the reasons why the fraction eluted quite late.

Structural parameters

From the data shown in Table I, an average molecular structural analysis was conducted using a method[3] that was developed by the author and is based on a Brown-Ladner method[5,6]. There are a lot of structural parameters in the average molecular structural analysis, but in this study, three parameters, number of total rings (Rt), number of aromatic rings (Ra), and number of units (M), are discussed since these are important parameters for the fused ring system in a molecule (Table II).

Rt is calculated by Eq. 1.

$$Rt = \frac{2 \cdot C + 2 - H - Ca}{2} \tag{1}$$

where C and Ca are numbers of total carbon and aromatic carbon, respectively. Since Rt/C, which describes the number of total rings per carbon (Eq.2), depends on the number of carbons due to the term 1/C, Rt', which is calculated by Eq. 3, was defined as a parameter for the number of total rings per carbon to evaluate the total ring number.

$$\frac{Rt}{C} = \frac{2 \cdot C + 2 - H - Ca}{2 \cdot C} \tag{2}$$

$$Rt' = \frac{Rt - 1}{C} = 1 - \frac{H/C + fa}{2} \tag{3}$$

where fa is the same as Ca/C. Ra/C and C/M were calculated from M and Ra estimated from the results of the average structural analysis. Here, C/M is the number of carbon per unit.

It is an unexpected result that the distributions of Rt' and Ra/C are nearly constant throughout the MW range (Figure 3). From Eq. 3, H/C+fa should be constant if Rt' is constant. When asphaltene is fractionated, H/C can readily be obtained from elemental analysis, but to directly measure fa is relatively difficult because much amount of sample was necessary for NMR analysis. Therefore, the relationship between them is very useful. It is widely known that H/C is a property which represent the aromaticity of heavy oil, i.e., the more aromatic oil has the less H/C. This correlation was applied for the structural analysis for heavy oil[7], where fa was determined from H/C, MW and density using a correlation diagram.

Ra/C displayed the same tendencies as Rt', which indicates that the parameters on the numbers of total and aromatic rings per carbon are constant throughout the MW range. The difference between Rt' and Ra/C represents the parameter on the naphthenic rings per carbon and is also constant.

Except for Fr. 6 in all the asphaltenes, C/M increased as the MW increased, but the degree of increase was less than the increase in MW. For example, the MW of Fr. 1 in AS-MY was 10.3 times larger than Fr. 4, but C/M was only 1.4 times. Figure 4 shows the correlations between MW and the formula weight of a unit calculated from C/M and H/C. The formula weight of a unit only increased from 400 to 1200 as the MW increased up to 8000. This figure indicates that Fr.1 consists of about 7 units since Fr. 4 is believed to consist of only one unit. Similarly, Frs. 1-2 for every asphaltene are thought to consist of several units.

Table II Avrage Structural Parameters

Fraction	Feed	1	2	3	4	5	6[*1]
Asphaltene from Iranian Light vacuum residue (AS-IL)							
C/M	50.7	63.8	63.3	52.9	45.9	36.8	136.4
Rt'	0.23	0.19	0.20	0.20	0.20	0.21	0.20
Ra/C	0.18	0.14	0.13	0.14	0.15	0.16	0.18
Ashaltene from Khafji vacuum residue (AS-KF)							
C/M	45.2	84.4	58.3	50.5	45.1	46.4	107.9
Rt'	0.20	0.20	0.19	0.20	0.21	0.22	0.17
Ra/C	0.15	0.12	0.13	0.12	0.14	0.14	0.17
Asphaltene from Maya vacuum residue (AS-MY)							
C/M	49.6	77.0	64.4	66.8	55.3	44.9	92.3
Rt'	0.20	0.18	0.19	0.16	0.18	0.20	0.21
Ra/C	0.17	0.12	0.14	0.13	0.14	0.14	0.17

*1 Fraction recovered from waste eluent.
M : Number of fused ring system in a molecule.
Rt : Number of total rings (aromatic and naphthenic rings).
Ra : Number of aromatic rings.
Rt' : Parameter on the number of total aromatic rings defined by eq. 3.

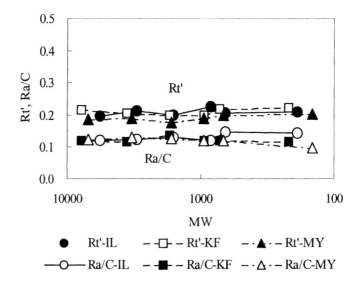

Figure 3 Parameter of the numbers of fused rings (Rt') and aromatic rings (Ra/C) for fractionated asphaltenes

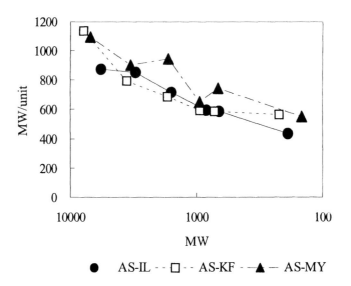

Figure 4 Formula weight per unit for fractionated asphaltenes

The MW of Fr. 5 determined by GPC was quite low, less than 250 (Table I), but the estimated MW from Figure 4 was between 400-600. The latter MW of asphaltene is more reasonable than the former MW since asphaltene is non-volatile and has a boiling point greater than 500°C.

Fr 6 had a MW similar to Fr. 2, but C/M was 1.5 – 2 times larger than Fr. 2, suggesting that Fr. 6 has a larger fused ring system than Frs. 1-5 and is a more coke-like fraction.

Molecular structural distribution

From the structural parameter distribution discussed above, the following results are plausible for the distribution of asphaltene molecules.

(1) Small molecules (MW of 400-600) consist of only one unit with relatively high fa.
(2) The fused ring system expands up to MW of about 1200 as the MW increases and the coupling of units occurs with a MW up to 1700.
(3) Above 1700 of a MW, polymerization and/or aggregation of several units are dominant. Therefore, most of structural parameters are constant in the range.

Those results are similar to those Kotlyar et al. reported on the maltene fration of Athabasca oil sand bitumen[8]. The maltene fractions were separated into two classes, MW<800 and MW>800, and fa drastically decreased as the MW increased in the former fraction, while the structural parameters changed slightly in the latter fration, which is a common behavior for maltene and asphaltene from vacuum residues.

Conclusion

Three kinds of asphaltenes from the vacuum residues of Iranian Light, Khafji, and Maya were separated into 5 fractions by a preparative GPC, followed by an average structural analysis to estimate the structural parameter distribution.

As the MW increased up to 1700, the H/C atomic ratio increased and nearly remained constant above 1700, while fa decreased. At higher MW, all parameters, including H/C and fa, remained similar. The number of total rings, which is related to H/C + fa, was almost constant throughout the observed range of MW as well as the parameters for the number of aromatic rings. N/C and S/C were also constant. C/M, the number of carbons per unit increased from 400 to 1200 as the MW increased. Summarizing the results, the molecular structure is classified into 3 types according to MW. The first is low MW molecules that consist of a single unit. The second is moderate MW molecules that consist of 1-2 units. The expansion of unit size is considerable on these two types. The last is high MW molecules that consist of several units by polymerization or aggregation with no further enlargement of a unit size.

Acknowledgment

This study was supported by an International Joint Research Grant Program from New Energy and Industrial Technology Development Organization (NEDO) and was conducted in collaboration between the National Institute of Advanced Industrial Science and Technology (AIST), the Central Research Laboratory, Idemitsu Kosan Co., Ltd., and the Argonne National Laboratory, USA.

References

1. Neurock M., Nigam A., Trauth D., Klein M. T., *Chem. Eng. Sci.*. **1994**, *49*, 4153-4177.
2. Calemma, V., Iwanski, P., Nali, M., Scotti, R., Montanari, L., *Energy & Fuels*, **1995**, *9*, 225-230.
3. Sato S., *J. Jpn Petrol. Inst.*, **1997**, *40*, 46-51.
4. Sato S., Takanohashi T., Tanaka R., *Prepr. Am. Chem. Soc., Div. Fuel Chem.*, **2001**, *46-2*, 353-354.
5. Brown, J. K., Ladner, W. R., *Fuel*, **1960**, *39*, 87-96.
6. Hirsch, E., Altgelt, K. H., *Anal. Chem*, **1970**, *42*, 1330-1339
7. Krevelen, D. W., Chermin, H. A. G., Schuyer, J., *Fuel* **1957**, *36*, 313-320
8. Kotlyar L. S., Morat C., Ripmeester, J. A., *Fuel*, **1991**, *70*, 90-94.

Chapter 6

Construction of Chemical Structural Model of Arabian Asphaltene and Its Density Simulation

Satoru Murata[1], Koh Kidena[2], and Masakatsu Nomura[2,*]

[1]Department of Industrial Arts and Crafts, Takaoka National College, 180 Futagami-machi, Takaoka, Toyama 933–8588, Japan
[2]Department of Applied Chemistry, Faculty of Engineering, Osaka University 2–1 Yamada-oka, Suita, Osaka 565–0871, Japan

In this study, the authors conducted structural analysis of Arabian asphaltene by ^{1}H-/^{13}C-NMR, oxidative degradation, and gel permeation chromatograph. Based on these data, the authors constructed structural model, whose molecular formula and molecular weight were $C_{490}H_{528}N_4O_5S_{15}$ and 7034.43 Da, respectively. The model constructed showed a fairly good agreement with analytical data such as elemental composition, distribution of hydrogen and carbon, average molecular weight, and so on. The authors also conducted computer simulation of physical density of the model and obtained the value of 1.07 g/cm^3, which was slightly lower than the value measured for Arabian asphaltene.

The authors have been studying the modeling of molecular structure of heavy hydrocarbons such as coal, solvent refined coal, and so on, because these heavy hydrocarbons consist of very complicated mixtures of relatively higher molecular weight hydrocarbons. In order to convert these materials into more useful and smaller molecular-weight-fractions, it is believed indispensable to know the details of chemical structures of these materials such as aromatic ring sizes and bridge bondings. For this purpose, measurements of molecular weight of original substances, identification of the bridge bonds of aromatic clusters and quantitative analysis of carbon functionalities of heavy hydrocarbons are in need. Molecular weight distribution of a complicated mixture of hydrocarbons is now determined by the use of LD/MS (laser desorption / mass spectrometry), although the more heavy portions and/or associated clusters can not be detected even by this method. As for the bridge bonds connecting aromatic nuclei, which are thought to govern the reactivity of heavy hydrocarbons, they have been studied by many researchers: For coal Stock et al. applied ruthenium ion catalyzed oxidation reaction to clarify these chemical structures (1) and Strausz et al. conducted the same reaction for the asphaltene derived from oil sand bitumen (2). The latter group published its model structure in 1992 and the former in 1993 based on their results. As for the quantitative analysis of carbon functionalities, Snape et al. (3) published the successful application of ^{13}C NMR spectroscopy with SPE/MAS (single pulse excitation / magic angle spinning) method. Using the data such as bridge bonds and carbon functionalities the authors obtained by themselves, they published the detailed model structure of Zao Zhung coal in 1998 (4).

However, in spite of these continuing efforts of many researchers, there are few means to examine the comprehensiveness of the model structures proposed so far. One candidate seems to be the estimation of physical density of proposed model structures using computer simulation method, however, this simulation needed an expensive software and high-performance computer, i.e. workstation, at the beginning of 1990's. Recently, such calculation could be easily attained due to the rapid growth of computer science. Several researchers still are proposing model structures of these heavy hydrocarbons without referring to the spacious arrangement of their chemical structures. This is the reason why the authors intended to present this article.

Construction of Chemical Structural Model of Asphaltene

Experiments

Samples

Vacuum residue (VR) of Arabian mixture (80 % Arabian Light and 20 % Arabian Medium) was provided by Nippon Oil Corporation. Asphaltene sample was obtained as a pentane insoluble fraction by Soxleht extraction of VR sample. The elemental composition of the asphaltene was as follows: C 83.7 %; H 7.5 %; N 0.84 %; S 6.8 %; O 1.11 %; Ni 0.012 %; V 0.038 %.

NMR measurements

NMR analyses were conducted by a JEOL JNM-GSX-400 spectrometer operating at 400 MHz for ^1H and 100 MHz for ^{13}C. The NMR samples were prepared by mixing approximately 100 mg of the sample with 1 mL of $CDCl_3$; tetramerthylsilane (TMS) was used as an internal standard. The quantitative ^{13}C-NMR measurements were attained by adding a relaxation agent, chromium trisacetylacetonate $[Cr(acac)_3]$ in inverse-gated decoupling system with a pulse delay of 5 s, acquisition time of 1.0 s and pulse with 3.3 µs.

GPC measurements

The measurement with a gel permeation chromatograph (GPC) for the estimation of molecular weight distribution of the asphaltene sample was conducted using a Shimadzu LC-6AD type pump and a Shimadzu SPD-10A type ultraviolet detector (wavelength = 270 nm). Tetrahydrofuran (THF) was used as an eluant with a flow rate of 1 mL/min. The column used in the measurements was a Shodex KF-806M, the packing materials of which is the mixture of several types of polystyrene gels with various types of particle size distribution. A series of standard polystyrene (purchased from Shodex) was adopted for the calibration of the relationship between the molecular weight and elution time.

RICO reaction

The RICO reaction was performed by stirring the mixture of the asphaltene sample (1 g), H_2O (30 mL), CCl_4 (20 mL), CH_3CN (20 mL), $NaIO_4$ (15 g), and $RuCl_3 \cdot nH_2O$ (40 mg) at 40 °C for 24 h. During the reaction, the resulting CO_2 was flowed through $CaCl_2$ and ascarite containing tubes with N_2 gas stream. The amount of CO_2 formed was determined from the weight increase of ascarite. The details of the workup procedure have been given elsewhere (5). Brief description of oxidation procedure is given bellow: For determination of lower carboxylic acids (from HCOOH to C_4H_9COOH), the reaction mixture was made basic by NaOH(aq) after the filtration procedure, followed by ion chromatographic analysis. For the analysis of other products, an oxidation mixture from another run was filtered and the precipitate was washed with CH₂Cl₂ (DCM) and water. The aqueous and organic phases were separated at first, the aqueous phase being further extracted with DCM. The aqueous phase was evaporated at 40 °C to dryness because the extraction of aqueous phase with solvent was found to fail in getting complete recovery of polycarboyxlic acids produced. Addition of 50 mL of ether solvent to the precipitate can recover polycarboxylic acids. Both the ether- and DCM-soluble portions were methyl-esterified by an ethereal solution of diazomethane (DAM). Esterified acids from both organic and aqueous phases were also subjected to vacuum distillation at 200 °C to determine the amount of heavy product. GC, GC/MS, elemental, and GPC analyses of the products were carried out as described previously.

Results

NMR analysis of the asphaltene

^{1}H- and ^{13}C-NMR spectra of the asphaltene were measured, the latter one being shown in Figure 1. Hydrogen was classified into four groups. Types and their chemical shift range of each group were as follows: aromatic hydrogen (H_{ar}), 9-6 ppm; hydrogen attached to carbon α to aromatic rings ($H_α$), 5-2 ppm; hydrogen attached to CH_n carbon β (or further positions, in case of CH and CH_2) to aromatic rings ($H_β$), 2-1 ppm; hydrogen attached to CH_3 carbon γ or further positions to aromatic rings ($H_γ$), 1-0 ppm. Concentration of each hydrogen class was determined based on integral of peak areas. Carbon types and chemical shift of ^{13}C-NMR region are as follows: methyl carbon (CH_3), 24-10 ppm; methylene or methine carbon (CH_2, CH), 38.5-28.5 ppm; naphthenic carbon (naphthenic), 60-38.5 ppm and 28.5-24 ppm; aromatic carbon ($C_{ar}=f_a$, carbon aromaticity), 160-109 ppm. Carbon types in aromatic region and equations for their calculation are as follows by referring to the elemental analysis data and ^{1}H-NMR data: protonated aromatic carbon (C_{arH}) = H_{ar} x (H/C); alkylated aromatic carbon (C_{arC}) = (H/C) x $H_α$ / (H_{al}/C_{al}); bridgehead and inner aromatic carbons (C_{arj}) = area (133-109 ppm) – C_{arH}; other aromatic carbon (attached to heteroatom etc.) = C_{ar} – (C_{arH} + C_{arC} + C_{arj}). Here, H/C is atomic ratio of hydrogen to carbon, determined by elemental analysis, and H_{al}/C_{al} means H/C ratio in the aliphatic portion obtained from NMR and elemental analysis data. Table 1 summarizes the structural parameters. Carbon aromatic value (f_a) was found to be 0.57. The parameter, χ_b, (6-7) which

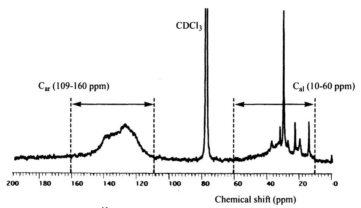

Figure 1. ^{13}C-NMR Spectrum of the Asphaltene

Table 1. Distribution of Hydrogen and Carbon in the Asphaltene.

Distribution of hydrogen $^{a)}$			
H_{ar}	H_{α}	H_{β}	H_{γ}
16.0	21.8	46.4	15.8

Distribution of carbon $^{b)}$								
C_{others}	C_{arC}	C_{arj}	C_{arH}	naphthenic	$-CH_2-$	$-CH_3$	f_a $^{c)}$	χ_b $^{d)}$
8.7	11.3	19.4	17.2	15.2	19.3	8.9	0.56	0.33

a) H_{ar}, H_{α}, H_{β}, and H_{γ} mean hydrogens attached to aromatic carbons and carbons at α-, β-, and γ-positions to aromatic rings, respectively.

b) C_{others}: total aromatic carbons – (C_{arC} + C_{arj} + C_{arH}), C_{arC}: alkyl groups-bearing aromatic carbons, C_{arj}: inner or bridgehead aromatic carbons, C_{arH}: tertiary aromatic carbons, naphthenic: carbons in hydrogenated aromatic rings, CH_2: methylenic carbons, and CH_3: methyl carbons.

c) f_a: fraction of aromatic carbons per total carbons.

d) χ_b: fraction of inner or bridgehead aromatic carbons per total aromatic carbons.

correlates with an average size of polycyclic aromatic units, was 0.33, this value corresponding to that of three or four ring-aromatic compounds.

RICO reaction of the asphaltene

We conducted RICO (ruthenium ion catalyzed oxidation) reaction of Arabian asphaltene to obtain structural information. Ruthenium tetroxide, RuO_4, is a strong oxidant and has a unique property, it having the nature to attack unsaturated carbons such as sp^2 or sp, preferentially. For example, alkylbenzenes, diarylalkanes, partially hydrogenated aromatics, polycyclic aromatic hydrocarbons were oxidized to aliphatic and aromatic carboxylic acids as shown in the following equations.

Therefore, the structure and the distribution of carboxylic acids produced from the reaction might give us valuable information concerning their chemical structure. The results are summarized in Tables 2 and 3, and Figure 3. The oxidation products were divided into 5 fractions, CO_2, lower aliphatic monocarboxylic acids, aqueous NaOH-soluble fractions (mainly consisted of aliphatic and aromatic carboxylic acids with two- or more carboxyl groups), DCM-soluble fractions, and insoluble fractions, summation of the yields of

Table 2. Carbon Balance of RICO of the Asphaltene

Fraction	Carbon recovery (mol / 100 mol of carbon in asphaltene)
Insolubles	4.4
Organic phase (DCM solubles)	24.5
Aqueous phase (NaOHaq solubles)	26.7
Lower monoacids (C_1-C_5)	9.3
CO_2	27.5
Total	92.4

Table 3. Yields of Aliphatic Carboxylic Acids from RICO of the Asphaltene.

Products (Yield / mol per 100 mol of C in asphaltene)

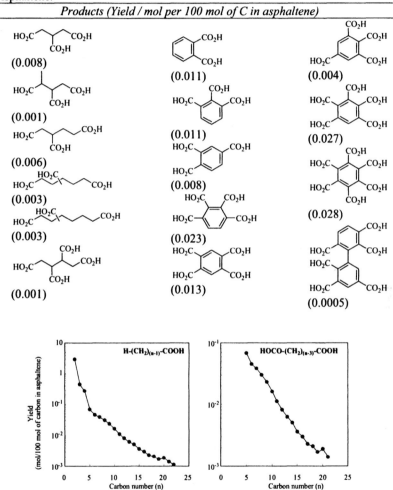

Figure 3. Yields of aliphatic mono- and diacids from RICO of the Asphaltene.

these fractions reaching more than 92 %. The value for carbon recovery seems to be enough for quantitative discussion. Yields of aliphatic mono- and diacids from RICO of the asphaltene are summarized in Figure 3. The longer the aliphatic chains in the acids were, the lower their yields were, these results indicating that methyl group is most important side chain in the asphaltene

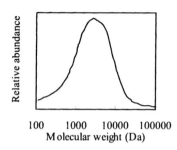

Figure 4. Gel Permeation Chromatogram of the Asphaltene

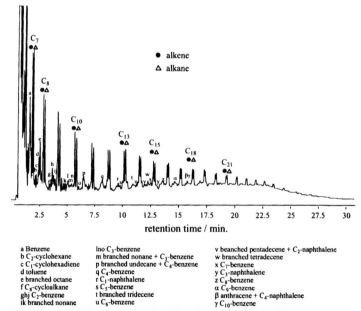

a Benzene	lno C_3-benzene	v beanched pentadecene + C_2-naphthalene
b C_2-cyclohexane	m branched nonane + C_3-benzene	w branched tetradecene
c C_1-cyclohexadiene	p branched undecane + C_4-benzene	x C_7-benzene
d toluene	q C_4-benzene	y C_3-naphthalene
e branched octane	r C_1-naphthalene	z C_8-benzene
f C_8-cycloalkane	s C_5-benzene	α C_9-benzene
ghj C_2-benzene	t branched tridecene	β anthracene + C_4-naphthalene
ik branched nonane	u C_6-benzene	γ C_{10}-benzene

Figure 5. Gas Chromatograph of Pyrolysis Products of the Asphaltene
(670 °C, 3 sec)

molecule. RICO of the asphaltene sample also afforded several aliphatic and aromatic polycarboxylic acids as shown in Table 3. The presence of these acids indicates that asphaltene structure contains variety of naphthenic and polycyclic aromatic moieties. The presence of benzene hexacarboxylic- and biphenyl hexacarboxylic acid in the oxidation products may indicate that there were considerable amounts of highly condensed aromatic compounds such as coronene in the asphaltene.

Molecular weight and pyrolytic properties

GPC analysis of the asphaltene was conducted, the resulting chromatogram being shown in Figure 4. Chromatogram of the asphaltene showed unimodal shape and ranged from 100 to 100,000 Da. Maximum abundance of molecular weight and number-average molecular weight were found to be 2.04 x 10^3 Da and 2.11x10^3 Da, respectively. Some researchers reported that average-size of molecular weight of petroleum- and oil sand bitumen-derived asphaltenes were found between 1500 and 5000 (8-9). Our results showed good agreement with these values reported.

Figure 5 shows the gas chromatogram of rapid pyrolysis products. About 44 % of the feed was pyrolyzed to afford pairs of alkane and alkene up to C_{27}, alkylated derivatives of benzene, naphthalene, and anthracene.

Construction of Structural Model of the Asphaltene

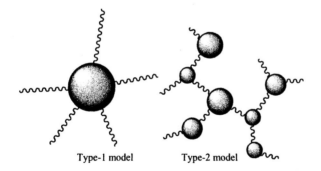

Type-1 model   Type-2 model

⬤ polycyclic aromatic moieties with partially hydrogenated parts

〰〰 alkyl side chains and bridges

Figure 6. Two Concepts for Model of Asphaltene Molecules.

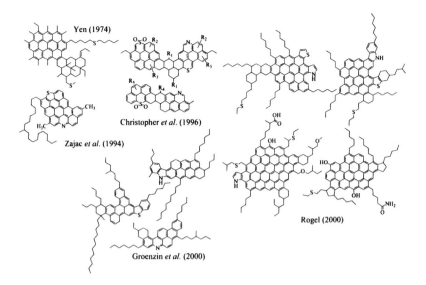

Figure 7. Condensed Aromatic Cluster Type Asphaltene Model

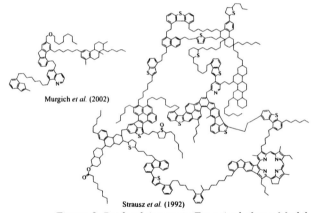

Figure 8. Bridged Aromatic Type Asphaltene Model

Based on the structural information mentioned above, the structural model for Arabian asphaltene was constructed. Before construction, the authors should determine their strategy for construction. Gray (10) had proposed two concepts for structural model of petroleum asphaltene: one is condensed aromatic cluster model (Figure 6, type-1) and the other is bridged aromatic model (type-2). The former has a polycondensed and huge aromatic cluster with 2-4 rings of naphthene and several alkyl side chains per one molecule, on the other hand, the latter consisted of several aromatic clusters with small or medium ring sizes connected by alkylene and thioether bridges. Figures 7 and 8 show the models of asphaltene proposed by Yen (11), Zajac (12), Groenzin (13), Rogel (14-15), Strausz (2), Murgich (16), which could be classified into these two categories. Gray concluded that concept of bridged aromatic model is most consistent with the results of thermal cracking of asphaltene, where asphaltene could afford a wide range of products. As far as the model structure of asphaltene, the authors agree with Gray's thought based on the following experimental results: (i) RICO reaction of asphaltene afforded a wide range of carboxylic acids along with minor amount of insoluble materials which may come from large aromatic clusters; (ii) an average size of polycyclic aromatic compounds was found to be around three or four; and (iii) rapid pyrolysis of the asphaltene afforded considerable amounts of volatile fractions, which contained a series of aliphatic and aromatic hydrocarbons. Therefore, the contribution of condensed aromatic cluster type model is considered as minor one. According to this estimation, the authors constructed structural model of Arabian asphaltene as the bridged aromatic model.

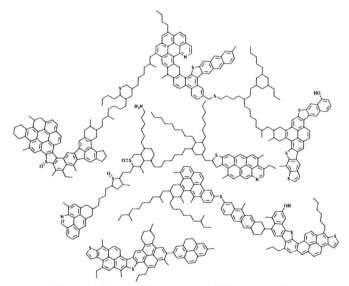

Figure 9. Chemical Structure Model for the Asphaltene.

Based on the structural analyses, the authors constructed the model of the asphaltene by employing the following assumption:

(a) Type of structural model is a bridged aromatic type.

(b) Distributions of carbons and hydrogens are determined based on the results of NMR analysis.

(c) Average ring size is around three or four based on NMR analysis.

(d) Molecular weight of the model ranges from 1,000 to 3,000 based on GPC analysis.

(e) Number of molecules is four within average molecular weight.

(f) Functional groups of sulfur in the asphaltene are thiophene, sulfoxide, and thioether, based on XPS analysis (17).

(g) Numbers of carbon, hydrogen, nitrogen, sulfur, and oxygen atoms are calculated based on elemental analysis and molecular weight assumed.

As to the assumption (e), the authors constructed four molecules with different molecular weights, because only one molecule with 2,000 Da of molecular weight (average value of molecular weight based on GPC analysis) could hardly reflect the experimental data, therefore, four is the possible least

Table 5. Comparison of the Data for the Constructed Model and the Asphaltene Sample

Sample or model	Elemental composition (wt%)					H/C
	C	H	N	O	S	
Measured value	83.7	7.5	0.84	1.11	6.8	1.07
Model	83.7	7.6	0.80	1.10	6.8	1.08

Sample or model	Hydrogen distribution (mol%)			
	H_{ar}	H_α	H_β	H_γ
Measured value	16.0	21.8	46.4	15.8
Model	16.7	23.3	44.1	15.9

Sample or model	Carbon distribution (mol%)							f_a
	C_{others}	C_{arC}	C_{arj}	C_{arH}	naphthenic	$-CH_2-$	$-CH_3$	
Measured value	8.7	11.3	19.4	17.2	15.2	19.3	8.9	0.566
Model	8.6	10.4	18.9	17.9	15.7	19.5	9.0	0.558

numbers of molecules for simplification. Since details for the procedure of construction of the model was reported in the literature (4), the authors will describe it briefly in the present paper. At first, a set of polycyclic aromatic hydrocarbons was selected based on the results of NMR and pyrolytic analyses, then, heteroatoms such as thiophene, sulfoxide, thioether, amino group, and pyridine were added to the above, and finally, naphthenic rings, alkyl side chains, and alkylene bridges were added based on NMR and RICO analysis. Thus, the authors obtained the structure shown in Figure 9. Molecular formula and molecular weight were $C_{490}H_{528}N_4O_5S_{15}$ and 7034.43 Da, respectively. Table 5 shows the comparison of structural features of the model with those of the real asphaltene, this indicating that the model well reflects the experimental data.

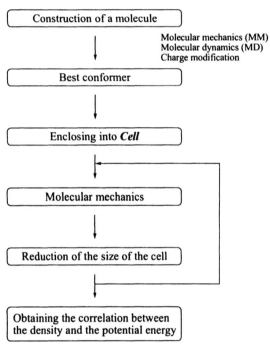

Figure 10. Outline for the simulation method of Physical Density of Asphaltene Model

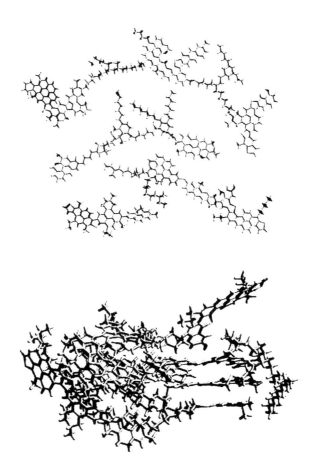

Figure 11. Three-dimensional Images of the Asphaltene, after construction and optimization by MM calculation (top) and after 10 ps MD followed by MM calculation (bottom) (for clarification, hydrogen atoms were eliminated).

Density Simulation of Asphaltene Model

Method for Calculation

CAMD studies were carried out using Cerius^2 (Accelrys Inc., ver. 3.6) on a SGI workstation (Indigo-II). Details of the methods were reported elsewhere (18-20). Figure 10 shows the procedure to estimate the density of model molecules. After inputting the model molecules on the software, the authors carried out the calculation of molecular mechanics and charge modification to optimize the potential energy of the model molecules until rms (root mean square) became less than 1.0 kcal/mol. The model was then submitted to molecular dynamics calculation for 50 ps at 300 K by quenched dynamics method to search more stable conformations. The conformation having the lowest potential energy was extracted. When the periodic boundary conditions are applied for calculation, this model molecule is enclosed in a cell. The density of the system is calculated from both the volume of this cell and the weight of all the atoms in the cell. When the model molecule was enclosed in the cell, potential energy of the system increased because the repulsive forces are caused by the contact of atoms among cells. Molecular mechanics calculation was then carried out to obtain a more stable conformation by considering the effect of surrounding molecules in the periodic boundary conditions until rms became less than 1.0 kcal/mol. In the molecular mechanics calculation, the inter- and intramolecular potential energies were optimized toward low potential energy by changing the conformation of the model molecule. Here, the size of the cell was decreased so that both the density and the potential energy of the system decreased. Molecular mechanics calculation was carried out in order to reduce the potential energy of the system until rms became less than 1.0 kcal/mol. This sequence was repeated for about 20-30 times. Through the cycles of this sequence, the potential energy of the system decreases with accompanying the increase of the density. The potential energy turned to increase via a certain minimum point where the authors have defined the density of the energy minimum as the true density of the system.

Results and Discussion

Figure 11 shows three-dimensional images of the asphaltene model optimized by Cerius^2. After optimization by only MM calculation, the structure was similar to two-dimensional image shown in Figure 8 (top of Figure 11), while after MD calculation, the conformation changed drastically and some aromatic stacking was observed (bottom of Figure 11). During MD calculation, potential energy decreased (from 1000 kcal/mol to 650 kcal/mol), these indicating that aggregation of asphaltene proceeds easily and aromatic stacking plays an important role in asphaltene aggregation. The resulting correlation between physical density and potential energy, and three-dimensional image of the density optimized structure of the asphaltene are shown in Figures 12 and 13,

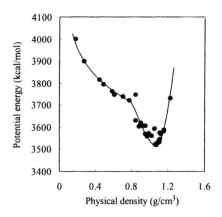

Figure 12. Correlation between Physical Density and Potential Energy Arabian Asphaltene

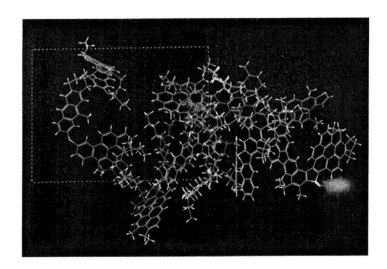

Figure 13. Density-optimized Model of the Asphaltene.

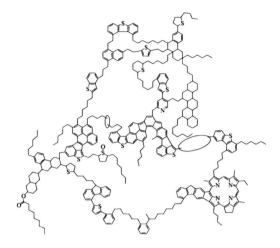

Figure 14. Modification of the Asphaltene Model Proposed by Strausz et al.[3]

respectively. Physical density of the model constructed was calculated to be 1.07 g/cm³.

The value obtained was slightly lower than the measured value, 1.20 g/cm³. The authors, here, tried to evaluate physical density of the asphaltene model proposed by Strausz et al, the model being also bridged aromatic type. By using same method for the Arabian asphaltene, the authors obtained the value of 0.98 g/cm³ as physical density of Strausz's model, this being considerably lower than the measured physical density of Athabasca asphaltene, 1.16 g/cm³ (21). The authors tried to modify Strausz's model. It has four loop structures as shown in Figure 8. In the previous report (20), loop structure tended to reduce physical density of the model, because loop structure disturbs effective folding of model structure. Therefore, the authors broke two bonds marked in Figure 14. Consequently, number of loop structures decreased from four to two. The physical density of the model modified was calculated to be 1.17 g/cm³. The results may indicate that molecular folding is one of the most important factors to calculate the density values as far as these calculation are applied for model structure.

By comparing the authors' model with Strausz's, the authors' contains some biaryl bonds and a biaryl thioether bridge, whose flexibility seems not to be larger than polymethylene bridges in Strausz's model. Therefore, effective compaction of Strausz's model modified could be attained. In the previous study (18), the authors conducted simulation of the following coal model compounds and obtained the conclusion that flexibility is one of the important factors to obtain compaction of model, i.e., monomethylene bridges disturb the molecular folding. This may partly due to the origin of low physical density of the model constructed. According to the data accumulated so far, employing smaller fragments for these models in accordance with the concept of aggregated structure of asphaltene could raise the density value. Therefore, above discussion should be limited within the model structures proposed.

n=1,2,3,4

Conclusion

In this study, the authors conducted structural analysis, construction of structural model, and density simulation of Arabian asphaltene. At first, the bridged aromatic type structure is favorable one compared with the condensed aromatic clusters type one. The model constructed according to the above concept showed a fairly good agreement with the analytical data such as elemental composition, distribution of hydrogen and carbon, average molecular weight, and so on., while its physical density estimated by computer simulation showed slightly lower than the value measured. The difference between the observed value and calculated value of density among two model structures was discussed by referring to the easiness of molecular folding.

References

1. Stock, L. M.; Tse, K.-T. *Fuel* **1983**, *62*, 974.
2. Mojelsky, T. W.; Ignasiak, T. M.; Frakman, Z.; McIntyre, D. D.; Lown, E. M.; Motgomery, D. S.; Strausz, O. P. *Energy Fuels* **1992**, *6*, 83.
3. Love, G. D.; Law, R. V.; Snape, C. E. *Energy Fuels* **1993**, *7*, 639.
4. Nomura, M.; Artok, L.; Murata, S.; Yamamoto, A.; Hama, H.; Gao, H.; Kidena, K. *Energy Fuels* **1998**, *12*, 512.
5. Artok, L.; Murata, S.; Nomura, M.; Satoh, T. *Energy Fuels* **1998**, *12*, 391.
6. Usually, the parameter, χ_b, a fraction of inner or bridgehead carbons to total aromatic carbons, increases with increase of the size of polycyclic aromatic units, for example, benzene (χ_b=0) < naphthalene (0.20) < anthracene and phenanthrene (0.29, 3 rings) < naphthacene (0.33, 4 rings) < pyrene (0.38, 4 rings) < benzpyrene and perylene (0.40, 5 rings) < coronene (0.50, 7 rings). Details were reported in the previous paper.[7]

7. Kidena, K.; Usui, K.; Murata, S.; Nomura, M.; Trisnaryanti, W. *J. Jpn. Petrol. Inst.* **2002**, *45*, 214.
8. Peramanu, S.; Pruden, B. B.; Rahimi, P. *Ind. Eng. Chem. Res.* **1999**, *38*, 3121.
9. Ancheyta, J.; Centeno, G.; Trejo, F.; Marroquin, G.; Garcia, J. A.; Tenorio, E.; Torres, A. *Energy Fuels* **2002**, *16*, 1121.
10. Gray, M. R. *Energy Fuels* **2003**, *17*, 1566.
11. Yen, T. F. *Prepr.-Am. Chem. Soc., Div. Pet. Chem.* **1972**, *17*(4), F102.
12. Zajac, G. W.; Sethi, N. K.; Joseph, J. T. *Scanning Microsc.* **1994**, *8*, 463.
13. Groenzin, H.; Mullins, O. C. *Energy Fuels* **2000**, *14*, 683.
14. Rogel, E. *Energy Fuels* **2000**, *14*, 566.
15. Rogel, E.; Carbognani, L. *Energy Fuels* **2003**, *17*, 378.
16. Murgich, J.; Merino-Garcia, D.; Andersen, S. I.; del Rio, J. M.; Galeana, C. L. *Langmuir* **2002**, *18*, 9080.
17. Artok, L.; Su, Y.; Hirose, Y.; Hosokawa, M.; Murata, S.; Nomura, M. *Energy Fuels* **1999**, *13*, 287.
18. Nakamura, K.; Murata, S.; Nomura, M. *Energy Fuels* **1993**, *7*, 347.
19. Murata, S.; Nomura, M.; Nakamura, K.; Kumagai, H.; Sanada, Y. *Energy Fuels* **1993**, *7*, 469.
20. Dong, T.; Murata, S.; Miura, M.; Nomura, M.; Nakamura, K. *Energy Fuels* **1993**, *7*, 1123.
21. Yarranton, H. W.; Masliah, J. H. *AICHE J.* **1996**, *41*, 3533.

Chapter 7

Molecular Constitution, Carbonization Reactivity, and Mesophase Development from FCC Decant Oil and Its Derivatives

Semih Eser, Guohua Wang, and Jennifer Clemons

Department of Energy and Geo–Environmental Engineering and Energy Institute, Pennsylvania State University, 101 Hosler Building, University Park, PA 16802

Carbonaceous mesophase development takes place during commercial delayed coking of FCC decant oil to produce needle coke. The graphitizability of the needle coke product depends strongly on the extent of mesophase development that indicates the microstructural ordering of large polynuclear aromatic species produced by carbonization. Laboratory carbonization of commercial FCC decant oil samples, their distillation fractions, hydrotreated feedstocks, and the charge to the coker (coker feed) showed significant variations in mesophase development under similar reaction conditions. Differences in molecular composition and initial carbonization reactivity of different samples were related to the observed differences in mesophase development. It was shown that rapid semi-coke formation in the gas oil fractions of the decant oil sample produced poor mesophase development. In contrast, hydrotreated gas oil fractions with a lower coking reactivity in the initial stages of carbonization produced semi-cokes that display a much higher degree of mesophase development. Also the coker feed samples, that consist of the high-boiling fraction of the decant oils with hydrotreated gas oils and the recycle from the delayed coker, produced a higher degree of mesophase development than that obtained from the corresponding decant oil samples carbonized alone. Blending hydrotreated streams into vacuum fractions of decant oils slows down the initial rate of semi-coke formation and improves the mesophase development, but decreases the semi-coke yield. In general, the abundance of pyrene and alkylated pyrenes in the feedstocks was observed to promote the desired mesophase development during carbonization. High concentrations of n-alkanes were found to be detrimental to mesophase development because of increased rates of carbonization.

Delayed coking converts petroleum heavy residua into liquid distillates and a carbon-rich byproduct, petroleum coke. Delayed coking produces a premium petroleum coke when a suitable feedstock is used under appropriate coking conditions [1]. This highly grahitizable carbon precursor, needle coke, derives its name from elongated domains of ordered microcrystalline structure that consists of large polyaromatic hydrocarbons. Needle coke has, therefore, a high degree of structural anisotropy and produces synthetic graphite with a very low coefficient of thermal expansion (CTE) upon graphitization heat treatment. Gaphite electrodes are used in electric-arc furnaces for recycling of scrap iron and steel.

The FCC decant oils, highly aromatic residua from Fluid Catalytic Cracking, constitute the primary feedstock for producing needle coke by delayed coking. The thermal conversion of this highly aromatic feedstock into solid carbon with highly anisotropic optical texture occurs through the carbonaceous mesophase development during liquid-phase carbonization [2]. Carbonaceous mesophase development requires the formation and parallel stacking of large, planar polynuclear aromatic hydrocarbon (PAH) that are called mesogens. Ordered domains of mesogens are deformed by the evolution of volatiles during the liquid-phase carbonization to produce the elongated domains of needle coke anisotropy as the viscosity of the liquid phase increases to form eventually a solid carbon [3,4].

Solubility fractionation has been used to characterize complex feedstocks such as decant oil [5]. A complex feedstock can be considered as being made up of three groups of components with regard to their roles in the carbonization [3]: (i). The low-molecular-weight fraction acts as a solvent during the period of carbonization and either vaporize from liquid phase or become incorporated into the mesophase by pyrolysis to mesogenic molecules; (ii). A higher-molecular-weight fraction is central to the formation of mesophase; (iii). A fraction insoluble in nearly all solvents gives isotropic carbon. Eser and Jenkins [6,7] reported that the asphaltene fraction (pentane-insoluble and toluene-soluble fraction) has a dominant effect on the mesophase development from a variety of feedstocks. These fractions have higher C/H ratio and higher propensity for carbonization to initiate the mesophase development.

Chromatographic fractionation of decant oils has been used to correlate the carbonization behavior with chemical composition [8,9]. It was found that a higher concentration of aliphatic compounds would be associated with the high reactivity of the feedstock particularly when large PAH are present in the feedstocks [9].

This study focuses on the analysis of the molecular composition of commercial decant oil samples and their derivatives obtained by distillation and hydrotreatment of selected fractions to relate mesophase development to molecular composition and carbonization reactivity of feedstocks. Samples of *coker feeds* that are directly introduced into the delayed coker from the botttom of the fractionator unit were also analyzed and carbonized along with the selected blends of decant oil derivatives.

Experimental Approach and Procedures

The samples used in this study include FCC decant oils (DO), gas oil fraction of decant oils (GO), hydrotreated gas oils (HYD), coker feeds (CF), and vacuum distillation bottoms (VTB) obtained from decant oils. Coker feed samples contain the high-boiling fraction of decant oils and the recycled heavy ends of the liquid products from delayed coking. Carbonization experiments were conducted in tubing bomb reactors (15 mL). About 4 grams of feedstock were accurately weighed and added into a reactor. The loaded reactor was then purged with nitrogen before plunged into a fluidized-sand bath. After reaction, the reactor was quenched in cold water.

Feedstock samples and selected liquid products from carbonization experiments were analyzed by Gas Chromatography/Mass Spectrometry (GC/MS). We used a standard mixture of PAHs (Supelco, Bellefonte, PA) to calibrate the instrument for quantitative measurements [8]. High Performance Liquid Chromatography with Photodiode Array detector (HPLC/PDA), Laser Desorption Mass Spectrometry (LD/MS) and Liquid Chromatography with dual mass spectrometers (LC/MS/MS) were also used to analyze the heavier aromatic compounds in the samples.

Two carbonization temperatures were used in the experiments. A relatively low severity carbonization was conducted at 450°C for 1 to 4.5 hours to monitor the changes at the initial stages of carbonization. The products were separated by dichloromethane extraction into semi-coke (DCM insoluble) and liquids (DCM soluble). The liquid products were extracted with pentane to separate asphaltenes (pentane insoluble) and maltenes. A higher temperature (500°C) was used in more severe carbonization experiments (for 4 h) to produce semi-coke in high yields for microscopic examination.. The semi-coke product was recovered as a whole piece and mounted in epoxy resin and polished for microscopic examination.

The extent of mesophase development in the liquid-phase carbonization of samples was measured in terms of an Optical Texture Index (OTI) after examination of the semi-coke samples under a polarized-light microscope (Nikon-Microphoto FXAII). We used a 1.1 mm X 1.1 mm mask and 10X objective lens to acquire the surface images. At least 250 images were examined for each semi-coke sample. The OTI of semi-coke was determined according to following equation [10]:

$$OTI=\sum f_i \cdot OTI_i$$

where f_i is the numerical fraction of individual texture types from microscopic analysis and OTI_i is the index value assigned to each texture type, as defined in Table 1.

Results and Discussion

Molecular Composition of Decant Oils and Their Derivatives

GC/MS Analysis

Decant oil has a complex molecular composition that consists primarily of 2- to 6-ring PAH and n-alkanes with carbon numbers from 8 to 32. GC/MS has proven useful to quantitatively analyze the molecular composition of the GC-amenable fraction up to 4-ring aromatics. Figure 1 shows the GC/MS total ion chromatogram of a decant oil sample, DO02-1, to represent a fingerprint of the complex molecular composition of decant oils. The overlapping peaks can be resolved using selected ion mass chromatograms (SIM) for the particular hydrocarbon species [8]. Table 2 list the concentrations of selected aromatic compounds determined by a GC/MS analysis for two different DO, CF, GO, and HYD samples. These three aromatic compounds ((naphthalene, phenanthrene, and pyrene) and their alkyl-substituted analogs (C1 to C3) constitute up to 20%wt of the feedstock samples and provide a useful mapping of the molecular composition of the aromatic compounds present in the samples [10].

In the two decant oil samples, naphthalene and its C1- to C3-analogs have similar concentrations, however, decant oil DO02-4 contains more pyrenes and less phenanthrenes than found in decant oil DO02-1. Gas oils have slightly higher concentrations of phenanthrenes and pyrenes compared to their parent decant oils. The differences observed in the degree of substitution of the aromatic compounds in the two decant oil samples are also reflected, in general,

Table 1. Definition of Optical Texture Index to characterize the optical textures of semi-coke samples

Type	Shape/ Size/	Index (OTI)
Mosaic	Isometric, <10 μm	1
Small Domain	Isomeric, 10-60 μm	5
Domain	Isometric/Irregular, >60 μm	50
Flow Domain	Elongated, >60 μm long, >10 μm wide	100

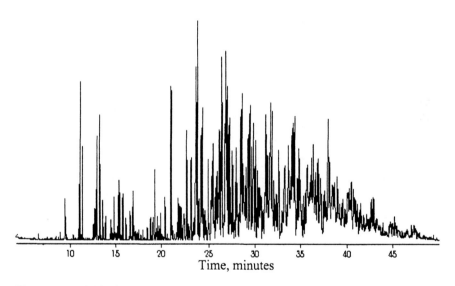

Figure 1. A GC/MS total ion chromatogram (TIC) for decant oil sample DO02-1.

Table 2. Selected GC -amenable aromatic compounds in feedstock samples (wt%).

	DO-02-1	DO-02-4	CF 02-1	CF 02-4	GO 02-1	GO 02-4	HYD 02-1	HYD 02-4
Naphthalene	0.08	0.09	0.04	0.04	0.00	0.00	0.01	0.00
C1-Naphthalene	0.42	0.35	0.09	0.11	0.20	0.20	0.04	0.03
C2-Naphthalene	0.78	0.67	0.07	0.11	0.43	0.42	0.09	0.04
C3-Naphthalene	0.52	0.50	0.00	0.00	0.33	0.19	0.10	0.03
Phenanthrene	0.58	0.68	2.20	2.15	0.62	0.88	0.13	0.11
C1-Phenanthrene	1.90	1.66	3.22	3.15	2.02	2.25	0.52	0.27
C2-Phenanthrene	2.39	2.18	3.02	2.56	2.69	3.12	1.27	0.70
C3-Phenanthrene	1.36	1.01	1.05	0.33	2.03	1.90	0.90	0.48
Pyrene	0.31	0.73	2.12	3.92	0.35	1.06	0.30	0.54
C1-Pyrene	0.69	1.19	2.48	3.56	0.84	1.58	0.80	0.94
C2-Pyrene	1.48	1.89	1.93	3.08	1.58	2.72	1.68	1.65
C3-Pyrene	0.28	0.90	0.56	2.58	0.41	0.94	0.57	0.49

in their GO derivatives. The coker feeds contain higher concentrations of phenanthrenes and pyrenes, particularly the unsubstituted compounds, presumably because of the recycle streams added to the heavy fractions of the decant oil. Among all the samples, CFs have the highest concentrations of GC-amenable aromatic compounds, reflecting again the significant influence of the recycle streams in enriching the more stable aromatic hydrocarbons, particularly the pyrenes, in the coker feeds. Hydrotreated derivatives show rather different composition from those of the respective gas oil samples. It appears that hydrogenation reduced naphthalenes and phenanthrenes proportionally more than the reduction observed in pyrenes. HYD02-4 sample shows, in general, more extensive hydrogenation of the parent GO than that observed in the 02-1 sample set. Note that the total detectable aromatics compounds in hydrotreated streams are much less in concentration than those in the corresponding gas oil samples, because partially hydrogenated naphthalenes, phenanthrenes, and pyrenes are not listed in the table.

In addition to PAHs, significant amount of normal alkanes are also present in the decant oil samples, particularly in DO02-1, as shown Figure 2. The normal alkanes present in DO range from C8 to C32 but most of the n-alkanes fall in the range of C20 to C26. Coker feeds contain lower concentrations of normal alkanes than the corresponding decant oils because of the higher thermal reactivity of n-alkanes than the GC-amenable aromatic compounds. Still, the coker feed CF02-1 has high concentrations of heavy n-alkanes resulting from the very high concentrations of n-alkanes in DO02-1 (Figure 2).

HPLC Analysis

Considering that a large portion of PAH are not resolved in GC/MS, to get further insight into the molecular composition of the feedstock samples High Pressure Liquid Chromatography with Photodiode Array detector (HPLC/PDA) and with tandem mass detectors (HPLC/MS/MS) s were used to qualitatively identify major PAHs in the samples. Figure 3 gives the HPLC/PDA chromatograms of CF02-1, CF02-4 and VTB02-4 at 254nm. The y-axis, AU, represents the quantity of absorption of uv light at 254 nm in absorption units; the x-axis is the time of elution from the HPLCcolumn. These chromatograms reveal very important features on heavy PAHs distribution that cannot be seen from GC/MS. The chromatograms in Figure 3 show a substantial difference in molecular composition of the aromatic compounds between CF02-1 and CF02-4. The second band of well-resolved peaks in the chromatogram of CF02-4 are assigned to five-ring PAH, indicating much higher concentrations of these compounds (e.g., benzopyrenes) in CF02-4 compared to those in CF02-1. The VTB02-4 shows a large hump of peaks in longer retention times that are assigned to six-ring and possibly larger PAH. The apparently lower concentrations of five-ring PAH in VTB suggest that these compounds are carried over by the recycle stream.

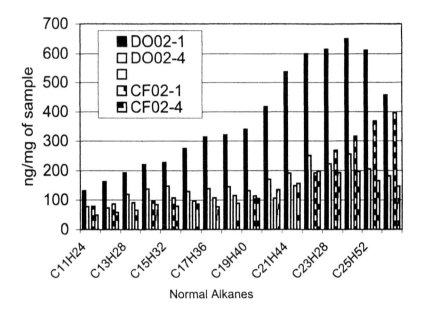

Figure 2. Distribution of n-alkanes in the decant oil and coker feed samples.

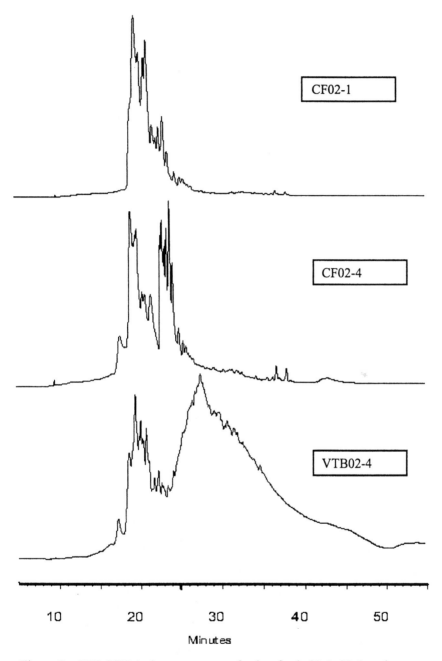

Figure 3. HPLC/PDA chromatograms of coker feeds 02-1, 02-4, and vacuum
bottoms 02-4.

Additional information on molecular composition of the coker feeds and the vacuum bottoms sample02-4 were obtained by HPLC/MS/MS analysis. Figure 4 shows the total ion chromatograms of these samples. There are very large difference between the molecular compositions of CF02-1 and CF02-4 in the heavier portions of these feedstocks. CF02-1, for example, contains much higher concentration of normal alkanes as shown in the initial region (retention time from 3 min to 8 min). This difference was not shown on the HPLC/PDA chromatograms (Figure 3), because PDA detector only detects compounds with conjugated π bonds; it does not detect alkanes. At longer retention time, CF02-1 shows significant concentrations of benzocarbazoles (polar compounds), and an unresolved envelope of very high molecular weight, or highly polar compounds that are virtually absent in CF02-4. In VTB02-4, the major peaks are shifted to longer retention times and more distinctive than in CF02-4; the six-ring PAH (e.g., benzoperylenes) constitute the majority of heavier aromatic compounds.

Mesophase Development from Carbonization of Decant Oils and Their Derivatives

Table 3 shows the optical texture data on the semi-cokes obtained from carbonization experiments at 500°C for 3 h. High Optical Texture Index values represent a higher degree of mesophase development, or more anisotropic optical texture in the semi-cokes as desired for premium needle coke. We see a significant variation among the degree of mesophase development in the two sets of feedstock samples, 02-1 and 02-4. DO02-1 produced a semi-coke with much lower extent of mesophase development (OTI: 69) while DO02-4 produced a more anisotropic coke with a high proportion of flow domain texture (OTI: 83). Coker feeds from both decant oils gave a higher degree of mesophase development than their parent decant oils, with OTIs 71 and 85 for DO02-1 and DO02-4, respectively. The vacuum bottoms (VTB) of DO02-4 produced an even more anisotropic coke (OTI: 88 of OTI). These differences in mesophase development can be related to the differences in the major hydrocarbon composition of these samples, as discussed in the previous section. Relatively low pyrene and alkylated pyrene concentrations along with high normal alkanes concentrations in DO02-1 and CF02-1 can explain the low degrees of mesophase development obtained from these samples compared to the more developed optical textures seen in the semi-cokes form DO02-4 and C02-4. The favorable effect of pyrenes and benzopyrenes is attributed to their hydrogen shuttling capabilities that control the rate of carbonization, or the rate of molecular growth processes and maintaining a sufficiently low viscosity that promote the mobility and alignment of mesogens for a high degree of mesophase development [7,10,11]. In contrast, the presence of high concentrations of n- alkanes can increase the rate of carbonization through facile formation of free radicals by the scission of C-C bonds in n-alkanes [10]. The

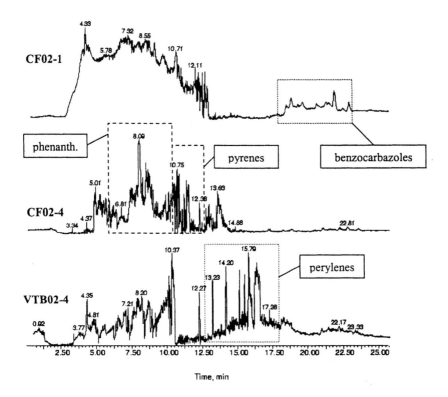

Figure 4. HPLC/MS/MS total ion chromatograms of the two coker feeds and VTB02-4.

Table 3. Optical textures of semi-cokes produced from the feedstock samples at 500°C for 4 h (6 h for HYD samples).

	Flow Domains	Domains	Small Domains	Mosaics	OTI
DO02-1	68	89	7	0	69
DO02-4	77	38	0	0	83
CF02-1	54	58	5	0	71
CF02-4	74	31	0	0	85
VTB02-4	105	35	0	0	88
HYD02-1	67	14	0	0	91
HYD02-4	53	8	0	0	93
GO02-1	32	68	23	10	51
GO02-4	40	80	12	7	58

abundant presence 5- and 6-ring PAH in VTB 02-4 with no n-alkanes present may account for the highly developed semi-coke texture obtained from this sample.

Hydrotreated samples gave high degrees of mesophase development (OTIs 91 and 93) presumably because of the low rate of carbonization due to the abundance of partially hydrogenated aromatics that act as hydrogen donors and low concentrations of n-alkanes. The complete conversion of hydrotreated samples to semi-coke took 6 h at 500°C compared to 4 h for the other samples. In contrast, gas oil samples produced the least developed semi-coke textures, that may be attributed to high n-alkane concentrations and high concentrations of alkylated phenanthrenes relative to pyrenes. Cracking of n-alkanes in the gas oil samples would lead to the formation of high molecular weight compounds and a rapid increase in the viscosity of the carbonization medium that inhibits mesophase development. The formation of low molecular weight alkanes from cracking reactions would also lead to the rapid precipitation of asphaltenes to form semi-coke in subsequent reactions.

Carbonization Reactivity: Semi-coke and Asphaltene Yields and Mesophase Development

Figure 5 shows the semi-coke and asphaltene yields from the coker feeds as a function of time at 450°C. By comparing sample CF02-1 with CF02-4, the rapid build-up (within 180 minutes) in asphaltenes in CF02-1 indicated the high reactivity of compounds in this coker feed. The short time window in the case of CF02-1 for the mesogens to stack and grow into large mesophase sphere and form elongated flow domain may explain the lower anisotropy of the semi-coke produced. The difference in reactivity between these two samples is consistent with their molecular composition difference: CF02-1 contains a higher concentration of normal-alkanes than CF02-4. Alkanes are the least thermally stable compounds in the carbonization environment that easily crack to form free radicals to serve as polymerization initiators. Higher concentration of alkanes in coker feed would result in a higher rate of carbonization that would lead to a lower degree of anisotropy in the semi-coke texture.

Figure 6 shows the semi-coke yields and mesophase content of the semi-cokes (measured by image analysis) of the heat-treated blends of VTB and HYD02-4 samples. Both semi-coke yields and mesophase contents are plotted on the same axis; lower value in the scale with the same symbol represents the mesophase content for a given sample. Similar to the trend observed with semi-coke yields, VTB alone produces the highest mesophase content at a given time and the mesophase contents of the solids produced from the binary blends decreases with the decreasing concentration of VTB. It is noteworthy that addition of HYD strongly inhibits the initiation of mesophase formation from VTB in the early stages.

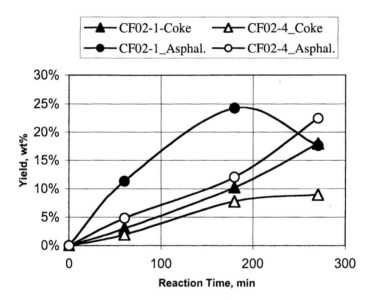

Figure 5. Semi-coke and asphaltene yields from carbonization of coker feeds at 450°C.

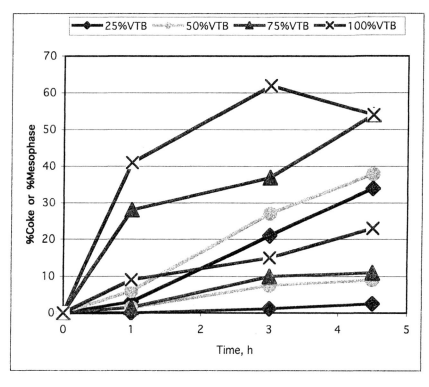

Figure 6. Percentages of semi-coke yield (higher values) and % mesophase (lower values) in the semi-coke from carbonization of VTB02-4 and VTB02-4/HYD02-4 blends at 450°C.

Figure 7 shows the OTI of the semi-cokes obtained from HYD, VTB and their blends at 500°C for 4 hours (6 hours for HYD alone). All the semi-cokes produced showed highly anisotropic textures with high OTI. The blending of HYD with VTB does not seem to have a significant effect on the resultant semi-coke texture, although 450°C experiments showed significant differences in the carbonization reactivities of the individual feeds and their blends. It should be noted however that addition of HYD to VTB requires longer carbonization times for complete conversion into anisotropic semi-cokes.

Conclusions

Laboratory carbonization of commercial FCC decant oil samples, their distillation fractions, hydrotreated feedstocks, and coker feeds was conducted to investigate the relationships between their molecular composition, carbonization reactivity and mesophase development. The examination of the resulting semi-coke texture by polarized-light microscopy showed significant variations in mesophase development under similar reaction conditions. The GC/MS, HPLC/PDA, and HPLC/MS/MS analyses demonstrated that significant differences exist in the molecular composition of the feedstock samples. Differences in molecular composition and initial carbonization reactivity of different samples were related to the observed differences in mesophase development.

It was shown that rapid semi-coke formation in the gas oil fractions of the decant oil sample produced poor mesophase development. In contrast, hydrotreated gas oil fractions with a lower coking reactivity in the initial stages of carbonization produced semi-cokes that display a much higher degree of mesophase development. Also the coker feed samples, that consist of the high-boiling fraction of the decant oils with hydrotreated gas oils and the recycle from the delayed coker, produced a higher degree of mesophase development than that obtained from the corresponding decant oil samples when carbonized alone. Blending hydrotreated streams into vacuum fractions of decant oils slows down the initial rate of semi-coke formation and improves the mesophase development, but decreases the semi-coke yield. In a previous study, Rahimi, et al. [12] have also observed under hot-stage microscopy that ,in fact, addition of hydrogen donors delays the induction period for mesophase formation from Athabasca bitumen vacuum bottoms.

In general, the abundance of pyrene and alkylated pyrenes in the feedstocks was observed to promote the desired mesophase development during carbonization. Good hydrogen shuttling capability of pyrenes is considered to play a significant role in mediating the rate of carbonization and promote mesophase development through prolonged fluidity of he carbonization medium. Higher concentrations of n-alkanes , on the other hand, result in a

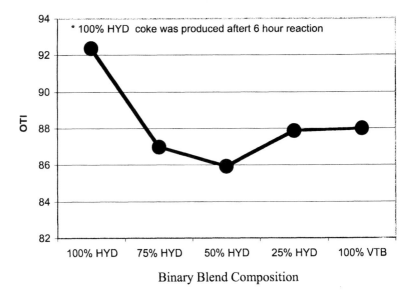

Figure 7. Optical Texture Indices for semi-coke produced from binary blends of hydrotreated gas oil (HYD02-4) and vacuum tower bottoms (VTB02-4) at 500°C, 4 hours.

Figure 8. Polarized-light micrograph of semi-coke from carbonization of HYD02-4 at 500°C for 6 h.

higher rate of carbonization that would lead to a lower degree of anisotropy in the semi-coke texture.

Acknowledgements

Financial support from Chicago Carbon Company and many helpful discussions with Robert Miller and John Bassett of Chicago Carbon Company are gratefully acknowledged.

References

1. Gary, J. H. and Handwerk, G. E., Petroleum Refining, Marcel Dekker, New York, 2001, pp. 71,72.

2. Brooks, J. D. and Taylor, G. H., *Carbon* 3, 185, 1965.

3. Mochida, I., Kudo, K., Fukuda, N., Takahashi, R. and Takeshita, K., *Carbon* 13,135,1975.

4. Mochida, I., Fujimoto, K. and Oyama, T. in Chemistry and Physics of Carbon, Vol 24, Edited by Thrower, P. A., Marcel Dekker Inc., New York, 1994, pp. 111-138.

5. Boenigk, W., Haenel, M.W. and Zander, M., *Fuel* **69**, 1226,1990.

6. Eser, S. and Jenkins, R., *Carbon* **27**, 877, 1989.

7. Eser, S. and Jenkins, R., *Carbon* **27**, 889, 1989.

8. Filley, R.M. and Eser, S., *Energy Fuels* **11**, 623, 1997.

9. Filley, R.M. and Eser, S., *Energy Fuels* **11**, 631, 1997.

10. Eser, S. in Supercarbon: Synthesis, Properties and Applications, Edited by Yoshimura, S. and Chang, R.P.H., Springer-Verlag, Berlin, 1998, p.147.

11. Murakami, K., Okumura, M., Yamamoto, M. and Sanada, Y., *Carbon* **34**, 187-192,1996.

12. Rahimi, P.M., Gentzis, T. and Fairbridge, C. and Khulbe, C., *Fuel Processing Technology* **60**,157-170, 1999.

Chapter 8

Vapor Liquid Equilibrium in Polycyclic Aromatic Compound Mixtures and in Coal Tars

Vahur Oja[1,2] and Eric M. Suuberg[1,*]

[1]Division of Engineering, Brown University, Providence, RI 02912
[2]Current address: Department of Chemical Technology,
Tallinn Technical University, Tallinn, Estonia

This paper presents the results of an experimental study on the vapor pressures of mixtures of high molecular weight aromatic compounds. Of particular interest are aromatics with heteroatomic content, especially nitrogen and oxygen. Such mixtures are of some interest in understanding the behavior of practical systems such as coal tars. The results demonstrate that significantly non-ideal behavior can be obtained in mixtures of high molecular weight (several hundred daltons), in the presence of even a single heteroatom in a molecule. On the other hand, the presence of heteroatomic species does not automatically mean that simple mixing rules will not be followed. Coal tar mixtures follow ideal-like behavior, even in the presence of significant hydroxyl content.

There exist very few data in the literature concerning the vapor-liquid equilibrium (VLE) behavior of mixtures of high molecular weight organics. This is because it is difficult to measure the vapor pressures of high molecular weight organics without their undergoing thermal degradation during measurement. This has typically required use of indirect vapor pressure measurement techniques. In this laboratory, one such technique, the Knudsen effusion method, has been used to measure the vapor pressures of pure aromatics (*1*), pure saccharides (*2*), and complex mixtures such as coal tars (*3*) and biomass tars (*4*). For the latter mixtures, a non-isothermal Knudsen effusion method has proven useful (*5*).

The present study addresses the problem of mixture behavior in complex high molecular weight organics, using the Knudsen effusion method, to measure vapor pressures. Again, this permits avoiding the problem of thermal degradation of compounds during measurement, the occurrence of which would of course invalidate the measurements. Vapor pressure data on such mixtures are of particular importance in the fields of fuels processing and combustion, in which vaporization of such species is a key step. The main focus of the present study is the question of whether high molecular weight polycyclic aromatic compounds (PAC) mixtures can be regarded as "ideal", and therefore, whether Raoult's law can describe their VLE behavior. It is of particular interest to examine how heteroatoms might affect the VLE behavior. The present results must be viewed as only a first step in addressing this complicated issue.

Experimental

The experiments involved preparing mixtures of known composition, and measuring their vapor pressures. In the case of many of the compounds of interest here, keeping temperatures low enough such that no thermal degradation would occur during measurements meant that the experiments were conducted under, or near, sublimation conditions. This might have had certain consequences for the results, as will be discussed below.

Equipment and Method

Vapor pressures of selected mixtures were indirectly determined, using the implementation of the Knudsen effusion method published earlier (*5*). Briefly, this method involves the measurement of the rate of effusion of a compound (or mixture components) through a pinhole leak in a capsule that is suspended from one arm of a continuously recording microbalance. These results are readily related to vapor pressures. The measurements were conducted in a high vacuum ($<10^{-7}$ torr), which permitted measurements of vapor pressure to as low as 10^{-6} torr. Because of the low vapor pressures that can be measured, the temperatures of measurement can be kept sufficiently low so as to prevent significant thermal degradation of the sample during the measurement.

Materials

The "pure" PAC materials examined here were reagent grade or best available purity materials, and were used as received with no further purification. The materials were in all cases of 98+% purity. Preparation of mixtures of PAC was accomplished using the so-called "quenching" method. This method involved measuring the desired amounts of two PAC compounds into a stainless steel capsule, under inert gas. After closing, the capsule was shaken, while the contents were heated to melting. "Instant" cooling was achieved by plunging the capsule into liquid nitrogen. It was assumed that this method provided "perfect" mixing of the components. No visible phase separation was observed in the samples prepared in this manner. It is fair to note that questions have been raised regarding the efficacy of this method, for preparing truly homogeneous crystalline mixtures (6). This will be considered further below.

The coal tar was prepared from a sample of Illinois No. 6 coal, obtained from the Argonne Premium Coal Sample Program (7). The tar was prepared by pyrolysis of the coal in a tube furnace, in inert gas at 700°C, and collected by washing the cold end of the reactor (at which the tars condensed) using tetrahydrofuran (THF). The THF solvent was fully removed from the tar sample by vacuum drying at 45°C. This procedure allowed capture of the tars with molecular weight greater than 150 daltons. Following preparation, the tar was separated on a preparative-scale gel permeation chromatography (GPC) column, using two styrene-divinylbenzene GPC columns in series. The separation solvent was THF. The separation was not a pure size separation, but depended somewhat on molecular properties. This is not important in the present work. The number average molecular weight of the tar was determined by vapor phase osmometry (VPO), in pyridine. The separated tar fraction was dried in vacuum at 50°C, to fully remove the THF solvent.

Results and Discussion

Mixtures of Pure Polycyclics

There can be a significant influence of heteroatom content on the vapor pressures of PAC, regardless of the small atomic fraction that the heteroatoms might represent in the molecule. This is illustrated by the results of Figure 1, in which the earlier reported (1) vapor pressures of pure pyrene and 1-hydroxypyrene are compared.

116

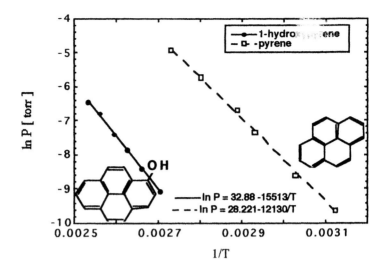

Figure 1. Comparison of vapor pressures of 1-hydroxypyrene and pyrene (1).

The formula weight of pyrene ($C_{16}H_{10}$) is 202.3, and this compound has a melting point of 429 K, whereas 1-hydroxypyrene ($C_{16}H_{10}O$) has a similar formula weight (218.3) and a melting point of around 453 K. The vapor pressures of these two compounds are very different, in the examined sublimation regime. Hence the influence of a small amount of heteroatomic content on the vapor pressures of such compounds cannot be ignored.

The present study began with an examination of the vapor pressures of various mixtures of pure PAC. Figure 2 shows results typical of those obtained in this study. The results are for an equimolar mixture of anthracene (MP=491 K) and perylene (MP=551 K). As might be concluded from the above melting points, the results in Figure 2 are, strictly speaking, for sublimation of the compounds. The predicted behavior, assuming that Raoult's Law is followed, is also shown in Figure 2. The calculation of the mixture vapor pressure in this case is assumed to follow:

$$P = x_1 P_1^0 + x_2 P_2^0$$

in which the x_i represent the component mole fractions and the P_i^0 represent the respective pure component vapor pressures. These pure component vapor pressures were determined as part of this study, and were generally in excellent agreement with values found in the literature (*1*).

While these data seem to suggest that Raoult's Law behavior might represent a fair approximation of the behavior, it is difficult to make a firm judgment of this, insofar as the disparity in pure component vapor pressures is so large that the lower molecular weight component (anthracene) dominates the behavior. Still, there do not appear to be any unexpectedly large deviations from ideality in what might be expected to represent a case of a mixture of materials with a high probability of ideal mixing behavior.

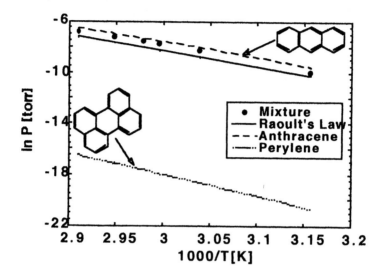

Figure 2. Vapor pressures of an equimolar mixture of anthracene and perylene.

On the other hand, Figure 3 shows that in a mixture of anthracene and benzofluorene, Raoult's Law is clearly not followed. This figure presents clear evidence of mixture non-ideality. The result might have been anticipated, since again, the mixture was well below the melting point of anthracene (491 K). Even though to the eye the mixture appeared homogeneous, it could well have been phase separated. In such a case, the two solid phases behave as thermodynamically separate entities, as far as the vapor pressures they exert. The tendency to approach Raoult's Law behavior at higher temperatures was observed here and in several other cases. This suggests that as the systems approach melting they begin to behave as more nearly ideal single phases.

What is important to note from Figure 3 is the direction of deviations from Raoult's Law. If two phases behave independently of one another, their combined vapor pressure is essentially the summation of the individual vapor pressures. The tendency to see deviations from pure, phase-separated behavior suggests that the compounds are interacting.

Keeping this in mind, the contrasting results of Figure 4, for a mixture of 1-hydroxypyrene and phenanthridine are considered. In this instance, the mixture

118

deviates in a direction of lower vapor pressure than the Raoult's Law prediction. This behavior cannot be explained by the possibility of phase separation.

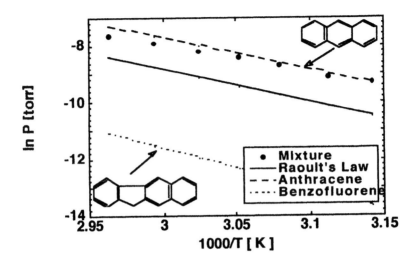

Figure 3. *Vapor pressure behavior of a mixture of 25 mol % anthracene and 75mol % bezofluorene.*

There is here strong evidence of a molecular interaction between the mixture partners. This could have perhaps been anticipated in light of the nitrogen base character of the phenanthridine and the acid (phenolic) nature of the 1-hydroxypyrene.

With these pure component and mixture results as a guide, the phase behavior of mixtures of PAC with coal tars was examined. A fraction of Illinois coal tar with a number average molecular weight of 270 daltons was mixed with pure compounds of similar molecular weight. It was estimated from elemental analysis that each "average" molecule of the tar contained about two hydroxyl groups, and the hydroxyl-rich character of this fraction was supported by its elution behavior in the GPC (its elution time was consistent with that of OH-rich species). The elemental analysis of this fraction showed that, by comparison, only every fifth tar molecule could contain nitrogen. In the case of the mixtures with tars, the phase of the measured phase was like that of a thick liquid.

Figure 5 displays the VLE behavior of a mixture of the Illinois coal tar and 1-hydroxypyrene. It is observed that the mixture of a phenolic compound with the coal tar produces what is very close to ideal mixture behavior. In this case, Raoult's Law was applied assuming that the tar could be treated as a single pseudo-component of molecular weight equal to the average molecular weight of the tar. The near ideal behavior is observed because the chemical nature of both components of the mixture is quite similar. The tar is believed to have a phenolic aromatic-rich structure. Hence addition of another phenolic compound to the mixture does not change the basic chemical nature of the mixture very much.

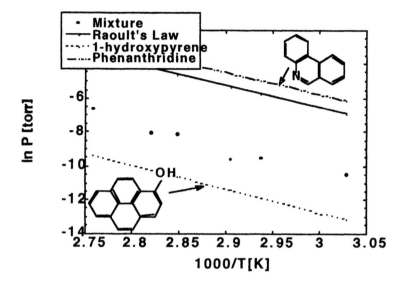

Figure 4. *Vapor pressure behavior of a mixture of 54 mol% 1-hydroxypyrene and 46 mol% phenanthridine.*

On the other hand, addition of phenanthridine to the coal tar produces a mixture (Figure 6) that shows a significantly lower vapor pressure than predicted by Raoult's Law, and consistent with the behavior observed in Figure 4. While coal tar itself contains a significant amount of nitrogen in the form of pyridinic structures, the numbers of such structures in the tar is quite small in comparison to the amount of oxygen (the latter present to a significant extent as hydroxyls. For example, the N/C elemental ratio in the coal tar was found to be 0.014, very similar to that in the parent coal (0.015). The estimate of O/C ratio (by difference) under the same conditions is 0.13, nearly an order of magnitude greater. Hence the phenolic character of the tar itself is expected to dominate the pyridinic character. Addition of a nitrogen compound permits the formation of a significant amount of new acid (phenolic) - base (nitrogen) interaction.

The implication of the above results is that the behavior of nitrogen compounds in the coal tar mixtures cannot be assumed to necessarily follow ideal mixture rules. This would have consequences for such things as the relative volatility of nitrogen containing tar species in comparison to species that do not contain nitrogen. There is some slight tendency in this direction, based upon preliminary results.

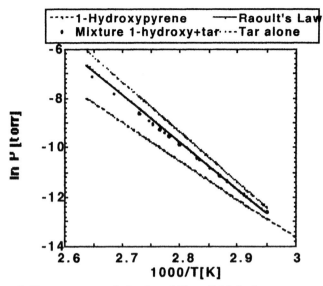

Figure 5. *Vapor pressure behavior of 50 mol% 1-hydroxypyrene with 50 mol% Illinois No. 6 coal tar.*

Conclusions

This study has examined, in a very preliminary way, the question of whether mixtures of high molecular weight aromatics can be treated as ideal, for the purposes of vapor pressure estimation. The evidence suggests that under certain conditions, ideal mixture behavior is approached, even if the mixtures contain a significant amount of heteroatomic species. On the other hand, the presence of heteroatoms can demonstrably lead to non-ideal mixture behavior. This was seen to be particularly the case when the mixtures contain heteroatomic species capable of forming strong specific interactions, i.e., nitrogen bases and hydroxyls.

Conclusions

This study has examined, in a very preliminary way, the question of whether mixtures of high molecular weight aromatics can be treated as ideal, for the purposes of vapor pressure estimation. The evidence suggests that under certain conditions, ideal mixture behavior is approached, even if the mixtures contain a significant amount of heteroatomic species. On the other hand, the presence of heteroatoms can demonstrably lead to non-ideal mixture behavior. This was seen to be particularly the case when the mixtures contain

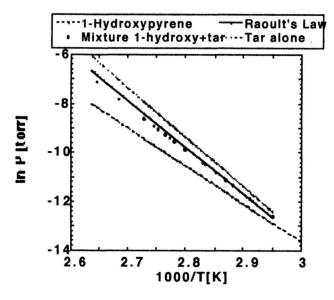

Figure 5. *Vapor pressure behavior of 50 mol% 1-hydroxypyrene with 50 mol% Illinois No. 6 coal tar.*

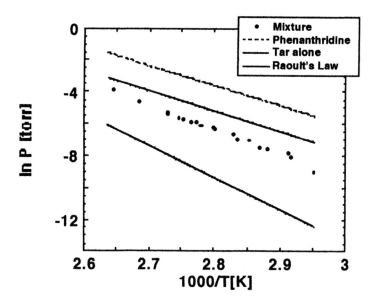

Figure 6. *Vapor pressure behavior of phenanthridine (35 mol %), and Illinois coal tar (65 mol%).*

122

heteroatomic species capable of forming strong specific interactions, i.e., nitrogen bases and hydroxyls.

Acknowledgement

The financial support of the U.S. Department of Energy, under grant DE-FG22-92PC-92544, is gratefully acknowledged. In addition, Brown University support for the Superfund group is also acknowledged.

References

1. Oja, V.; Suuberg, E.M. *Jl of Chem. Eng. Data*, **1998**, *43*, 486.
2. Oja, V.; Suuberg, E.M. *Jl of Chem. Eng. Data*, **1999**, *44*, 26.
3. Oja, V.; Suuberg, E.M. *Energy and Fuels* **1998**, *12*, 1313.
4. Suuberg, E.M.; Milosavljevic, I.; Oja, V. *Proc. Combustion Institute*, **1996**, *26*, 1515.
5. Oja, V.; Suuberg, E.M. *Anal. Chem.*. **1997**, *69*, 4619.
6. DeKruif, C.G.; van Genderen, C.G.; Bink, J.C.W.; Oonk, H.A. J., *J. Chem. Thermodyn.* **1981**, *13*, 1081.
7. Vorres, K. *Energy and Fuels*, **1990**, *4*, 420.

Chapter 9

Processing and Characterization of Shale Oil

New Methods for Characterization of the Effluent from Thermal Cracking of Oil Scale and the Products from Microactivity Testing of Its Atmospheric Residue

Yevgenia Briker[*], Zbigniew Ring, and Cecilia Sin

[1]National Center for Upgrading Technology,
[1]Oil Patch Drive, Devon, Alberta T9C 1A8, Canada

Most of the crude oils used for fuel production consist predominantly of hydrocarbons and contain only relatively small amounts of sulphur, oxygen, and nitrogen. Shale oils, in contrast, can contain large amounts of organic sulphur and/or oxygen compounds. Therefore, during processing of shale oils, in addition to making fuels, there may be opportunities to extract valuable chemicals. Knowledge of the physical and chemical properties of these shale oils is very important for the development of technologies necessary to fully utilize them. In this chapter we discuss detailed chemical characterization of the material derived from severe thermal treatment of oil shale. We also examine the potential for increasing its utilization through application of the fluid catalytic cracking (FCC) process to its atmospheric residue fraction boiling between 320 and 750°C.

The scope of this study includes analytical characterization thermally cracked shale oil for the purpose of developing a conceptual proc design for upgrading of this material to make marketable products and to enable estimation of their value. This characterization was carried out to assess the feasibility of extracting valuable phenolic compounds as the first processing step, and utilizing the remaining hydrocarbon matrix for fuel production. With respect to assessing further processability of shale oil, a simulated fluid catalytic cracking of the atmospheric residue fraction was carried out and the value of the effluent from this process was also assessed.

Fluid catalytic cracking (FCC) is the most important heavy gas oil conversion process. FCC generates products ranging from light olefins through to naphtha (the key product), and from light cycle oil to heavy cycle oil. The FCC micro-activity test (MAT) has been commonly used to simulate the FCC commercial process at the laboratory scale. In this chapter, we discuss the analytical characterization work performed to explore the possibility of applying the FCC process (as represented by MAT) to the atmospheric residue derived from thermally cracked shale oil.

Shale oil rich in organic oxygen is a source of fuel oil and synthesis gas, but it could also be a source of a number of valuable chemicals such as antiseptic oil for wood impregnation, electrode coke, rubber softeners, casting binders, etc. Phenols found in such shale oils are used as feedstocks for epoxy and other resins, and as glue compounds, rubber modifiers, synthetic tanning agents, etc. The most valuable phenols are found in the 200-360°C boiling range. Therefore, there is interest in exploring the behavior of 360°C+ shale oil residue in the cracking environment, particularly in terms of the production of phenols in the useful boiling range.

From the analytical point of view, the main challenge of this study was in the development of a characterization methodology to analyze mono- and di-hydric phenols in the shale oil. The added difficulty was that MAT products are inherently produced in very small amounts (<2mL) and have a very wide boiling range.

A characterization procedure described by Černý et al. (1) was adopted as the basis of the new separation procedure developed at the National Centre for Upgrading Technology for quantification and identification of various fractions derived from the catalytically cracked shale oil residue. The extrography procedure was replaced with an open column chromatography, and the elution solvents were used in the order reported earlier (2). A new way of calculating the material balances was proposed for the full boiling-range material.

Experimental

The thermally cracked shale oil and the liquid product from MAT processing of its atmospheric residue were both analyzed for the types and the amounts of various phenols.

Analysis of Thermally Cracked Shale Oil

In the first part of the study, a sample of the thermally cracked shale oil was distilled into five fractions following the ASTM D2892 method: light naphtha (IBP – 150°C), heavy naphtha (150 – 200°C), atmospheric distillate (200-360°C), vacuum distillate (360-520°C) and residue (520°C+).

The light naphtha fraction did not contain any phenols (the boiling point of the lightest phenol – hydroxybenzene – is 182°C). At the other end of the boiling point range, the residue fraction (520°C+) was not amenable to most of the GC-based analytical methods used in this study. Therefore, light naphtha and residue fractions were removed first and disregarded in the subsequent search for phenols. By removing the light naphtha, we also expected to minimize mass losses due to handling during further analysis and, therefore, to improve the mass balance calculations that followed. Some physical properties of the shale oil and its fractions are presented in Table I. The simulated distillation curves of all the distillation fractions are shown in Figure 1.

The original thermally cracked shale oil contained high concentration of organic oxygen-containing compounds. Further processing of the distillation fractions relied on the removal of these compounds from the hydrocarbon matrix by a combination of water and caustic washings. The schematic shown in Figure 2 is a rough indication of the envisaged commercial process.

Table I. Physical Properties of Shale Oil and the Distillation Cuts

	Shale Oil	Light Naphtha (LN)	Heavy Naphtha (HN)	Atmospheric Distillate (D1)	Vacuum Distillate (D2)	Residue (R)
BP (°C)		IBP-150	150-200	200-360	360-520	520+
Yield (wt%)	100	10.9	9.4	30.4	20.2	29.1
Density (g/ml)	1.018	0.7761	0.8225	0.9296	1.042	1.4349
Carbon (wt%)	75.2	84.26	84.54	84.24	82.44	55.94
Hydrogen (wt%)	9.2	12.9	12.7	11.13	9.22	4.58
Oxygen (wt%)	5.32	1.68	1.96	4.29	7.12	6.73

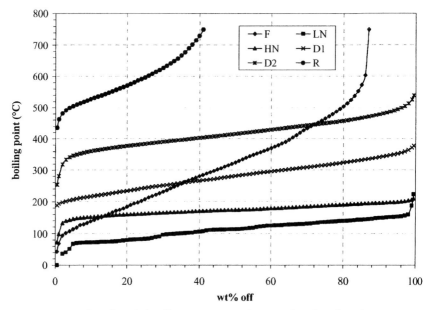

Figure 1. Simulated distillation curves of the shale oil and its fractions
(Reproduced with permission from *ACS Fuel Chemistry Preprints* **2003,** *48(1),*
56. Copyright 2003 by the authors.)

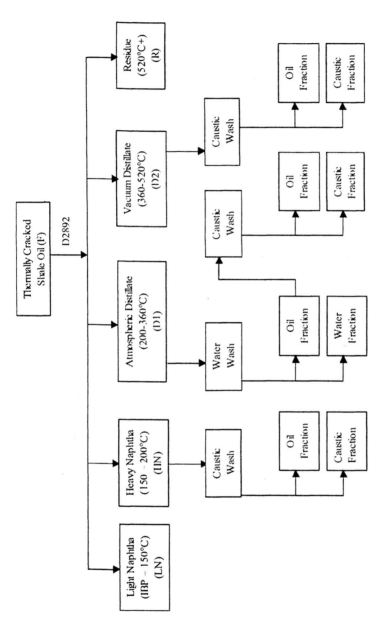

Figure 2. Processing of the shale oil and its fractions

The three middle fractions, HN, D1 and D2, were washed to extract oxygen-containing compounds. Since the most valuable phenols boil in the 200-360°C temperature range, more emphasis was put on possible fractionation of various oxygen-containing compounds during the extraction of Fraction D1. It was washed with water first to explore the least expansive option. We expected that the water wash would remove some phenolic compounds. The rafinate from the water wash of Fraction D1, and the other two fractions, HN and D2, were subjected to a caustic wash (12.0wt% of NaOH in water). The three extracts from the caustic wash were designated as '**Acids**' and the corresponding rafinates were designated as '**Neutrals&Bases**'.

Each caustic wash used 100 mL of the solution and was applied to each sample in three equal subsequent portions. Each organic rafinate (Neutrals&Bases) from the caustic wash was separated from the corresponding extract and washed three times with 100 mL of water (again in three equal portions). The aqueous extracts from the water wash of organic rafinates were added to the caustic extract. The organic rafinates that resulted from this water wash of the D1 and D2 fractions were diluted with dichloromethane and dried with Drierite. After drying, the solvent was removed by evaporation to recover the extracted material, which was subsequently quantified by weight. In order to prevent any loss of the HN material during the evaporation step, the dilution and drying steps were not applied to the HN-derived rafinate. In the case of Fraction D2, because of the high viscosity, the sample was dissolved in dichloromethane before the caustic wash. The less viscous Fractions HN and D1 were washed without dilution.

The caustic extracts derived from Fractions HN, D1 and D2, and containing sodium phenolates, were acidified with hydrochloric acid (acid was added drop wise to the phenolate solution while mixing until the pH reached 2) to recover phenols. The resulting solutions corresponding to Fractions D1 and D2 were re-extracted with 100 mL of dichloromethane (applied in three equal batches), dried with Drierite, and quantified by weight after removing the solvent with the help of a rotary evaporator. In the case of the HN-derived caustic extract, to prevent the loss of the sample during drying and rotary evaporation, after the extraction the phenolic fraction was carefully separated from the water layer but was not re-extracted with dichloromethane or dried with Drierite.

MAT Processing of Atmospheric Residue

In the second part of the study, the residual material derived from the shale oil was used as feed for MAT FCC simulation. The following conditions were used in this test: reactor temperature 490°C, commercial catalyst, contact time 30 sec, and catalyst/oil ratio 3.952. Selected properties of the feed and the total liquid product (TLP) are presented in Table II. The corresponding simulated distillation curves are shown in Figure 3.

Table II. Properties of Feed and TLP

Feed		TLP	
BP, °C	*%OFF*	*BP, °C*	*%OFF*
IBP (301.2°C)	0.5	IBP (84.5°C)	0.5
FBP (750.0°C)	99.5	FBP (587.5°C)	99.5
IBP-360°C	4.16	IBP-200°C	13.81
360-525°C	60.91	200-360°C	36.69
Density, g/mL	1.0660	Density, g/mL	1.0106

Source: Reproduced with permission from *ACS Fuel Chemistry Preprints* **2003**, *48(1)*, 56. Copyright 2003 by the authors.

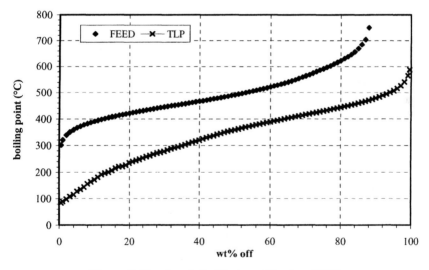

Figure 3. Simulated distillation of feed and TLP

Analysis of Products

The total liquid product from the MAT processing of the shale oil atmospheric residue and both the 'Acids' extracts and the 'Neutrals&Bases' rafinates from the washing of the HN, D1, and D2 distillation fractions were analyzed for phenol content using a liquid chromatography analytical procedure. This procedure was a modification of an extrography method described in the literature (1). The extrography column that was equipped with plunger pump and filled with basic alumina and silica in the original method was replaced with an open glass column (13.5-mm I.D. and 450-mm length) filled with silica gel. Silica gel from Fisher Scientific (28-200 mesh) was activated for 12 h at 140°C. Its activity was adjusted by the addition of 4wt% water. Whereas in the original method an analyzed sample was evenly distributed throughout the whole surface of the silica gel prior to extraction, in our modified method, the total amount of sample was applied on top of the column. The elution solvents were used in order of increasing polarity (2). Since the TLP and the heavy naphtha contained some of the material boiling below 200°C (13.81wt% and 95.6 wt%, respectively), the calculation of material balance using gravimetric determination, as in the original method, would not be suitable. Removal of solvents by vacuum rotary evaporator and vacuum dryer would result in the loss of light ends and poor mass balance. In our modified method, the solution of each fraction was reduced to a known volume (for example, 1mL) by using a mild rotary evaporation and a nitrogen gas sweep. These reduced solutions were further analyzed by gas chromatograph-flame ionization detector (GC-FID). The fractions were quantified by using an external calibration. The calibration standards were prepared previously from shale oil by the identical LC separation followed by complete removal of solvents and gravimetric quantification of separated fractions. From these fractions of calibration standards the solutions with known concentrations were prepared and then analyzed by the GC-FID with the same method as the analyzed fractions. The response factors determined for each calibration fraction were used to calculate the yields of the corresponding fractions separated from the analyzed samples. All fractions were quantified only from 200°C; therefore the slight loss of the light ends encountered during the nitrogen purge did not affect the results. To determine the 200°C cut point, the retention time calibration was performed by analyzing the normal paraffin standards by GC-FID at identical conditions. The elution sequence and fraction designations are shown in Table III.

Table III. Elution Sequence and Volumes for LC Separation

Number	Designation	Solvent	Volume of solvent, mL
Fraction 1	Saturates + Olefins	n-Hexane	120
Fraction 2	Aromatics	Toluene	100
Fraction 3	Esters+	Chloroform	80
Fraction 4	Monophenols	90%Chloroform/ 10%Ether	150
Fraction 5	Diphenols	85%THF/ 15%Methanol	60
Fraction 6	Polars	Methanol	40

Results and Discussion

Phenols in Shale Oil

The results of LC separations of the fractions derived from the caustic wash of the three middle distillation fractions of the shale oil are presented in Table IV.

The values for subtotals of 'Acids' and 'Neutrals&Bases' are obtained from the washing procedure gravimetrically. The values for subtotals of 200°C+ and 200°C- fractions for 'Acids' and 'Neutrals&Bases' are calculated from the simulated distillation results of the corresponding fractions. The total numbers in the bottom row reported as the "wt% of total "correspond to the D2892 distillation yields of the cuts reported in Table I.

The 'Acids' fraction contained mostly phenols. Furthermore, according to LC separation, Fraction 4 would consist mostly of mono-phenols and Fraction 5 would consist mostly of di-phenols. To confirm the phenolic nature of Fractions 4 and 5 separated from D1 and D2 'Acids', these fractions were subjected to GC-MS analysis with the library search option. Fractions 4 and 5 of HN "Acids" were too small and were not analyzed by GC-MS. Instead, the total fraction of the HN 'Acids' was analyzed by the GC-MS, and the peaks were verified by the MS library. The nature of compounds separated in Fractions 4 and 5 of D1, D2 and HN 'Neutrals&Bases' was unclear and had to be verified also by mass spectrometry.

Chromatograms of some of the GC-MS runs with the library search results are shown in figures 4,5 and 6.

The total ion chromatogram of HN 'Acids' (not shown here) displayed three major peaks of phenols in the 200°C- portion and a few small peaks of substituted phenols in the 200°C+ portion of the chromatogram.

Table IV. LC of the Shale Oil

	Yield (wt% of HN)	Yield (wt% of Total)	Yield (wt% of D1)	Yield (wt% of Total)	Yield (wt% of D2)	Yield (wt% of Total)
Neutrals+Bases						
200°C+						
Fraction No.						
1	1.7	0.2	34.9	10.6	3.8	0.8
2	0.3	0.0	25.0	7.6	36.4	7.4
3	0.2	0.0	1.8	0.5	6.2	1.3
4	0.7	0.1	6.5	2.0	15.0	3.0
5	0.7	0.1	2.5	0.8	5.2	1.1
6	0.0	0.0	0.6	0.2	2.1	0.3
Subtotal (200°C+)	3.6	0.4	71.3	21.7	68.7	13.9
200°C-						
Subtotal (200°C-)	90.9	8.5				
Subtotal (Neutrals+Bases)	94.5	8.9				
Acids						
200°C+						
Fraction No.						
1	0.0	0.0	2.3	0.7	0.1	0.0
2	0.1	0.0	3.4	1.0	4.8	1.0
3	0.0	0.0	2.7	0.8	2.3	0.5
4	0.5	0.1	9.1	2.8	12.5	2.5
5	0.1	0.0	4.8	1.5	9.1	1.9
6	0.0	0.0	0.5	0.1	2.5	0.4
Subtotal (200°C+)	0.7	0.1	22.8	6.9	31.3	6.3
200°C-						
Subtotal (200°C-)	4.8	0.4				
Subtotal (Acids)	5.5	0.5				
Subtotal (Water Wash)			5.9	1.8		
Total	100.0	9.4	100.0	30.4	100.0	20.2

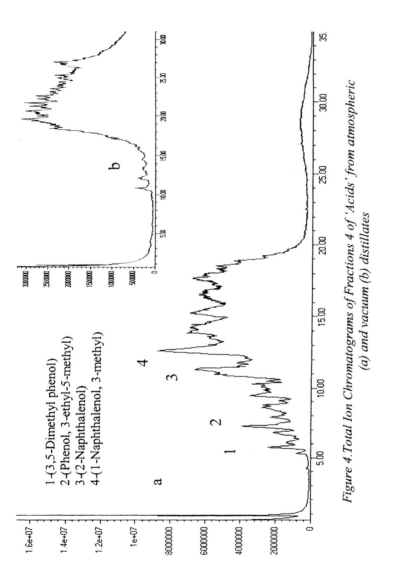

1-(3,5-Dimethyl phenol)
2-(Phenol, 3-ethyl-5-methyl)
3-(2-Naphthalenol)
4-(1-Naphthalenol, 3-methyl)

Figure 4. Total Ion Chromatograms of Fractions 4 of 'Acids' from atmospheric (a) and vacuum (b) distillates

Figure 4 shows the total ion chromatograms of Fractions 4 of 'Acids' from the D1 (a) and D2 (b) fractions of shale oil. According to the results of GC-MS library search, Fraction 4 of 'Acids' from atmospheric distillate (Figure 4a) consisted mostly of substituted phenols and naphthalenols. The chromatogram of Fraction 4 of 'Acids' of vacuum distillate (Figure 4b) was poorly resolved and, therefore, the results of the library search were ambiguous. However, based on good consistency of the LC separation and the results shown for Fractions HN, D1and D2, this fraction probably contained mostly the substituted mono-phenols boiling at higher temperatures.

The chromatograms of Fraction 5 of D1 and D2 'Acids' are shown in Figures 5(a) and 5(b).

These chromatograms show that the dominant compound types in Fraction 5 of 'Acids' were di-phenols. There is one particular di-phenol compound, phenol 2,5-bis (1,1-dimethylethyl)-4-methyl that is present in this fraction in a very large amount. There were no di-phenols found in the total 'Acids' of HN fraction.

The GC-MS analysis of Fractions 4 from the 'Neutrals&Bases' showed that the major compounds in these fractions were ketones (Figure 6).

The GC-MS analysis of Fractions 5 of D1 and D2 'Neutrals&Bases' did not show any presence of resorcinols. There was some carryover of phenol 2,5-bis (1,1-dimethylethyl)-4-methyl. Apparently this compound was present in this particular shale oil at a very high concentration. The carryover suggests that the number of caustic washes could be increased or higher concentration of caustic could be used to improve the separation.

Based on the above observations the following conclusions can be made about the concentration and the distribution of phenols in the shale oil.

From Table IV, the amounts of Fraction 4 in the 'Acids' of atmospheric distillate and vacuum distillate cuts were 2.8wt% and 2.5wt%, respectively, based on the total shale oil. These numbers total 5.3wt%, and according to GC-MS analysis, this entire amount can be attributed to mono-phenols. The HN 'Acids' constituted 0.5wt% of total oil with only 0.1wt% of the fraction boiling above 200°C. This entire amount belonged to Fraction 4, and consisted of a few mono-substituted homologues of phenol. The 200°C- portions of the HN 'Acids' and 'Neutrals&Bases' fractions were 4.8wt% and 90.9wt%, respectively, based on the HN fraction, or 0.4wt% and 8.5wt% based on the total shale oil. There were no phenols found in the 200°C- portion of 'Neutrals&Bases' but there were three major peaks of phenols found in the 200°C- portion of 'Acids'. If this entire amount (0.4wt%) is assumed to be mono-phenols, then the total amount of mono-phenols determined in the shale oil was 5.8wt%.

The amounts of di-phenols in the HN, D1, and D2 distillation fractions were 0.0wt%, 1.5wt% and 1.9wt%, respectively. These numbers total 3.4wt% of di-phenols based on total shale oil. Considering that 1.8wt% of water-washed D1 material consisted of di-phenols (due to their higher solubility in water), the total

135

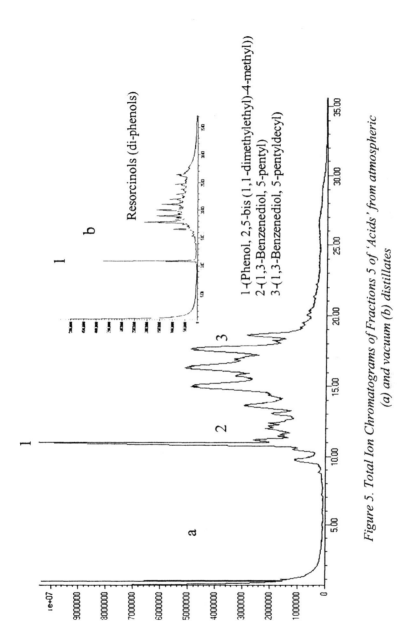

Figure 5. Total Ion Chromatograms of Fractions 5 of 'Acids' from atmospheric (a) and vacuum (b) distillates

Resorcinols (di-phenols)

1-(Phenol, 2,5-bis (1,1-dimethylethyl)-4-methyl))
2-(1,3-Benzenediol, 5-pentyl)
3-(1,3-Benzenediol, 5-pentyldecyl)

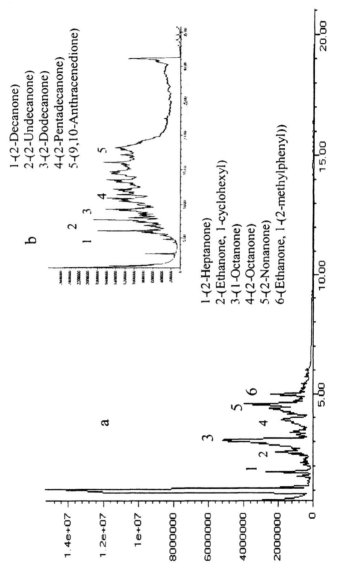

1-(2-Decanone)
2-(2-Undecanone)
3-(2-Dodecanone)
4-(2-Pentadecanone)
5-(9,10-Anthracenedione)

1-(2-Heptanone)
2-(Ethanone, 1-cyclohexyl)
3-(1-Octanone)
4-(2-Octanone)
5-(2-Nonanone)
6-(Ethanone, 1-(2-methylphenyl))

Figure 6. Total Ion Chromatograms of Fractions 4 of 'Neutrals&Bases' from atmospheric (a) and vacuum (b) distillates

amount of di-phenols in shale oil is 5.2wt%, bringing the total amount of phenols to at least 11.0wt% [some loss of phenol 2,5-bis (1,1-dimethylethyl)-4-methyl was due to carryover to Fraction 5 of 'Neutrals&Bases'].

A few small peaks of organic acids were detected on the GC-MS chromatograms of HN 'Acids' and of Fraction 4 of D1 'Acids'. Apparently they were present in shale oil in small amounts and were extracted with phenols by the caustic wash. This LC separation usually renders the organic acids in Fraction 4.

Phenols in MAT Product

The chromatogram of 'Acids' of D2 Fraction 4, shown in Figure 4b, was not well resolved. Apparently this fraction contains high-molecular-weight mono-phenols that do not have significant commercial value. As mentioned above, the residue fraction with the boiling range of 520°C+ was not amenable to GC-based methods used in this study. Since these two fractions (D2 and residue) represented 42.1wt% of the total shale oil, there was an interest in catalytically cracking this material to lower the molecular weight of phenols and produce a lower boiling range material containing more valuable phenols.

MAT processing of the material derived from the shale oil with IBP 301.2°C and FBP 750.0°C yielded a product with IBP 84.5°C and FBP 587.5°C. From the corresponding simulated distillation data, it was calculated that 13.81wt% of the material in the product was boiling below 200°C. The yield of the material boiling between 200-360°C was 36.69wt%, while there was only 4.1wt% of material boiling below 360°C in the feed (Table II). This temperature interval was important since it represented the fraction that might contain the phenolic compounds of interest, produced by cracking of heavy material.

The results of LC separation of TLP are presented in Table V, and the GC-FID chromatograms of Fractions 4 and 5 are shown in Figures 7 and 8. Each fraction from LC separation was analyzed by GC-FID and quantified in the interval of interest (between 200°C and 360°C) as described above. The retention times corresponding to these boiling points were determined from the retention time calibration and were 5.727 min and 17.482 min, respectively. In the bottom row of Table V, the results calculated for the 200°C to 360°C interval are balanced with the results calculated from simulated distillation data of the TLP shown in Table II.

According to the original method (1), and also confirmed by our analysis of shale oil, the bulk of the phenolic compounds are found in Fraction 4 (mono-phenols) and Fraction 5 (di-phenols).

Table V. LC of the TLP from Shale Oil

Fr. No	Solvent	Designation	LC/ GC-FID Recovery, wt%	200- 360°C SimDist wt%	IBP- 360°C SimDist wt%	360°C- 200°C FBP SimDist wt%	IBP- FBP Sum wt%
			200-360°C				
Fr.1	Hexane	Saturates	2.3				
Fr.2	Toluene	Aromatics	24.6				
Fr.3	Chloroform	Esters+ Chlor./	0.6				
Fr.4	Dieth.Ether	Monophenols	5.2				
	THF/						
Fr.5	Methanol	Diphenols	3.3				
Fr.6	Methanol	Polars	0.7				
Total			36.7	36.7	13.8	49.5	100.0

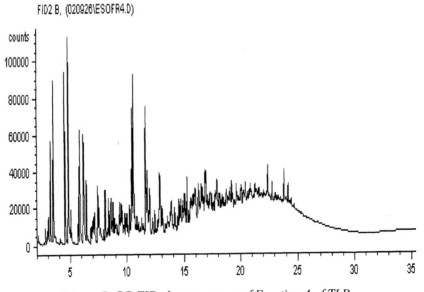

Figure 7. GC-FID chromatogram of Fraction 4 of TLP
(Reproduced with permission from *ACS Fuel Chemistry Preprints* **2003**, *48(1)*,
56. Copyright 2003 by the authors.)

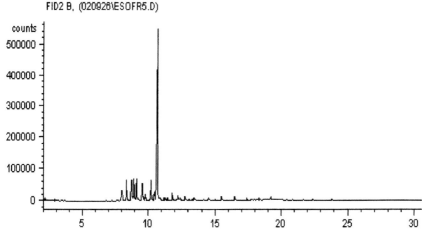

Figure 8. GC-FID chromatogram of Fraction 5 of TLP

(Reproduced with permission from *ACS Fuel Chemistry Preprints* **2003**, *48(1),*
56. Copyright 2003 by the authors.)

From Table V, the combined phenolic fraction (Fractions 4 and 5) was the
second largest (8.5wt%) after the aromatic Fraction 2 (24.6wt%). To confirm the
phenolic nature of Fractions 4 and 5, they were analyzed by GC-MS with the
library search option. The results are shown in Table VI.

Most of the peaks in Fraction 4 were identified as phenols with the largest
peaks representing orto-, meta-, para-cresols and 2,5-dimethyl phenol at 6.78
min, 6.86 min, 7.10 min, and 7.57 min, respectively. The rest of the
chromatogram represents the variety of di- and tri-methyl phenols. The
separation of the TLP was carried out without pre-washing due to a very small
yield of TLP (the usual MAT test yields about 2 mL of TLP). There were no
acids found in this fraction. Apparently, they decomposed during MAT
processing. The ketones detected in abundance in Fraction 4 of the
'Neutrals&Bases' of shale oil were also not present. Apparently, they also
decomposed during MAT processing. No clear assignments for peaks were given
by the GC-MS in Fraction 5 due to a very low concentration of individual
compounds. The largest peak was identified as phenol 2,5-bis (1,1-dimethyl)-4-
methyl. This is consistent with the previous findings and suggests that Fraction 5
consists mostly of di-phenols.

Conclusions

During the course of this study, water and caustic washing procedures were
tested with regards to their ability to extract phenolic compounds from three
distillation fractions of shale oil. This study showed that, when using a 12wt%
caustic solution and following the washing sequence specified above, the total
amount of phenols extracted from the three middle fractions was about 11.0wt%

Table VI. Peak Assignments for Fraction 4 of TLP

Peak RT,min	Assignation
5.70	phenol
6.78	phenol,2-methyl
6.86	phenol,2-methyl
7.10	phenol,3-methyl
7.57	phenol,2,5-dimethyl
8.03	phenol,2-ethyl
8.18	phenol,3,5-dimethyl
8.47	phenol,3-ethyl
8.60	phenol,2,3-dimethyl
8.84	phenol,2,4-dimethyl
8.99	phenol,2,3,6-trimethyl
9.32	phenol,2-ethyl-5-methyl
9.39	phenol,2-ethyl-6-methyl
9.47	phenol,3-(1-methylethyl)-
9.53	phenol,3-ethyl-5-methyl
9.77	phenol,2-(1-methylethyl)-
9.88	phenol,3,4,5-trimethyl
9.94	phenol,2,4,6-trimethyl
10.58	4-allylphenol
10.95	phenol, 3,5-diethyl
11.90	6-methyl-4-indanol
12.44	phenol, 4-(3-methyl-2-butenyl)-
13.06	1-naphthalenol
13.18	2-naphthalenol
14.33	1-naphthalenol,2-methyl
14.39	1-naphthalenol,3-methyl
14.64	1-naphthalenol,4-methyl
15.30	benzaldahyde, 2-chloro-6-fluoro-
15.48	1-naphthol, 5,7-dimethyl
16.37	2-hydroxy-1-naphthalenepropanol

Source: Reproduced with permission from *ACS Fuel Chemistry Preprints* **2003,** *48(1),* 56. Copyright 2003 by the authors.

based on the total shale oil sample. Out of this amount, 5.8wt% was identified as mono-phenols, and 5.2wt% was identified as di-phenols. One particular compound identified in Fraction 5 of the caustic extract was phenol 2,5-bis (1,1-dimethylethyl)-4-methyl, and it was present in significant amounts. Some amount of this compound was also carried over to Fraction 5 of the 'Neutrals&Bases' in an amount that was not identified and therefore not included as part of the total di-phenols in shale oil. The amounts of Fraction 5 of the 'Neutrals&Bases' in heavy naphtha, atmospheric and vacuum distillate fractions were 0.1wt%, 0.8wt%, and 1.1wt%, respectively, based on total shale oil. These total 2.0wt%. According to the analysis of the total ion chromatograms of Fractions 5, only part of this amount can be attributed to phenol 2,5-bis (1,1-dimethylethyl)-4-methyl, which is less than 2.0wt%. Good separation of phenols and ketones was achieved by the caustic wash, as confirmed by LC and GC-MS analysis.

Catalytic cracking of the residual material with density 1.066 g/mL derived from the shale oil resulted in the production of a lighter liquid product with the density 1.0106 g/mL, and also resulted in quite a significant shift in the boiling point range. The starting material boiling between 301.2°C and 750°C was converted to a lighter material boiling between 84.5°C and 587.5°C. The usefulness of this material in terms of phenolic compounds was verified by applying the modified LC separation technique and GC-MS analysis. The separation showed that as a result of catalytic cracking, 8.5wt% of phenols in the useful boiling range was produced.

A new procedure was developed for analyzing shale oil samples.

Acknowledgment

Partial funding for NCUT has been provided by the Canadian Program for Energy Research and Development (PERD), the Alberta Research Council (ARC) and the Alberta Energy Research Institute (AERI). This paper was partially funded by VKG of Estonia.

References

1. J. Černý, H. Pavlicová and V. Machovič, *"Compound-class fractionation of coal-derived liquids by extrography"*, FUEL, **1990**,Vol 69, 966-971.
2. M.M. Boduszynski, R.J. Hurtubise and H.F. Silver, Anal. Chem., **1982**, 54, 375.

Chapter 10

Hamaca Crude Oil Upgrading Using Formic Acid as Hydrogen Precursor under Steam-Injection Conditions

O. Delgado, C. Bolívar, C. Ovalles, and C. E. Scott*

Centro de Catálisis Petróleo y Petroquímica, Facultad de Ciencias, Universidad Central de Venezuela, Apartado Postal 47102, Los Chaguaramos, Caracas 1042A, Venezuela
*Corresponding author: cscott@ciens.ucv.ve

The effect of formic acid, as hydrogen precursor, in the upgrading of Hamaca extra heavy crude oil, was studied. Reactions with crude oil dispersed in sand (10 % by weight of oil), in a batch reactor at 280 °C and 88-109 Bar of N_2, and water, were carried out. These conditions were chosen in order to physically simulate those commonly found in a crude oil reservoir during production by steam injection. All reactions were carried out for 24 hours, and the upgraded oil was analyzed by SARA determinations, thermogravimetric analysis, sulfur and nitrogen content, and by gas chromatography (only saturate fractions). It was found, that decomposition of formic acid produced H_2 which is effectively used for upgrading Hamaca crude. When formic acid is used in conjunction with water, water acts as catalyst for thermal decomposition of formic acid, and the upgrading, measured in terms of SARA, sulphur and light fraction content, is more effective.

143

Introduction

Venezuela, a country traditionally associated with oil, currently has total crude reserves which amount to 221 billion barrels, of which 76 billion are proven reserves. Of the proven reserves 69% are heavy and extra-heavy crudes. Heavy and extra-heavy crudes can also be found in other countries in vast abundance, and for instance, the heavy oil and tar sands bitumen in Alberta Province of Canada and in the Orinoco Belt of Venezuela are each as large as the oil deposits in Saudi Arabia[1]. However, as compared to conventional crude oils, there are tremendous disadvantages in the recovery, transportation and refining of heavy crude oils because their densities and viscosities are much higher than the conventional ones. Venezuela's heavy oil viscosity at reservoir conditions is so high that the final recovery of some of them has been estimated to be around 5 to 15 %. The use of downhole upgrading has been proposed in order to enhance heavy crude oil recovery. Several methods, as deasphalting[2-5], visbreaking[6-9] underground hydrogenation[10-17] or *in situ* combustion[18-20], as well as hydrogen donor solvents treatments[21-23] have been proposed as possible enhanced recovery technologies. Also, the injection of hydrogen precursors downhole has been reported as a way of *in situ* partially upgrading of heavy crude oils. In fact, a 1992 patent[16] claims that *in situ* hydrogenation in a subterranean formation is readily performed by introducing a non gaseous hydrogen precursor into the oil bearing subterranean formation to enhance oil recovery. Formic acid and its organic and inorganic salts are among the hydrogen precursors proposed in the patent. It is reported that after treating a crude oil in a batch reactor with formic acid and MoS_2 (as catalysts), at 290 °C and 109 Bar, API gravity of the upgraded oil is increased and Asphaltene content is reduced, however, a detailed characterization of the upgraded product is not given.

In this work, a detailed study of the effect of formic acid in the upgrading of Hamaca extra-heavy oil, under steam injection conditions, was carried out. Laboratory physical simulations, using a batch reactor, at 280 °C and 88-109 Bar will be presented. The reactor was fed with a mixture of crude oil and sand (99 w % SiO_2) having 10 w % crude oil, together with formic acid, and water. We believe that this conditions more closely represent the actual situation found during crude oil production by steam stimulation processes.

Experimental

Upgrading reactions were performed in a stainless-steel 300 cm^3 batch reactor (Parr), without stirring. In a typical test, the reactor was fed with 55 g. of Hamaca crude oil (from the Orinoco Belt with the following properties: 9.1 °API Gravity, 14.2 % Carbon Conradson, viscosity (60 °C)= 9870±170 cP, 7500 ppm nitrogen 3.79 % sulfur, 450 ppm of vanadium and 102 ppm of Ni) and sand (silboca, 99 w% SiO$_2$, specific area < 1 m^2 g^{-1}) containing 10 % by weight of oil, 5 g. of formic acid and 5g. of distilled water. Experiments with no water added were also carried out. The reactor was heated at 4.2 ° C min^{-1}, to 280 ° C. Prior to heating the system was purged with nitrogen and pressurized up to 34 Bar. Final pressure, after heating, was around 88 Bar (for experiments without Formic Acid) and 109 Bar (for experiments with Formic Acid). All reactions were carried out for 24 hours. The reactor was then allowed to cool down to room temperature and oil was removed from sand by solvent extraction with dichloromethane.

Original Hamaca and Extracted oils were analyzed for sulfur and nitrogen, using a 9000NS sulfur and nitrogen analyzer from ANTEK. Thermogravimetric analysis were also carried out in a simultaneous differential techniques instrument SDT 2960 TA Instruments by heating the samples in nitrogen (10 °C min^{-1}, gas flow 60 ml min^{-1}). Saturates, aromatics, resins and asphaltene (SARA) separation and quantification was done in a Iatroscan MK-5 on samples separated on thin layer chromatography and detected by a hydrogen flame ionization system. Finally, saturate fractions were isolated by eluting a deasphalted sample with n-hexane on an alumina column, and analyzed by gas chromatography in a Perkin Elmer autosystem XL equipped with a FID and a capillary column (6M * 0.53 mm HT5 Simd)

Results and discussion.

SARA analysis results are presented in Table I. It is found that treating the Hamaca crude oil in the presence of water (W) produces a decrease in asphalthene and resins (13.9 and 35.5 % conversions respectively), with the corresponding increase in the aromatics and saturates fractions. This is in agreement with previous report[24] in which a reduction of the asphaltenes and resins fractions, and the corresponding increase of aromatics and saturates is found after treating heavy crude oils in the presence of steam.

When the crude oil was treated in presence of formic acid (FA), a higher degree of asphaltene conversion was obtained (23.1 %), while resins are converted to a lesser extend (22.2%). Again aromatics increases, but in this case, saturates are close to its content in the original oil. When formic acid was used in conjunction with water (FA+W) the best results are obtained. Thus, asphaltenes are reduced in 29.2 % and Resins in 44.5 %. Obviously, formic acid is more efficient in upgrading the oil when used together with water.

Table I. SARA fractions for untreated and upgraded Hamaca crude oil.[a]

	Saturates/ wt%	Aromatics/ wt%	Resins/ wt%	Asphaltenes/ wt%
HAMACA	(8.0 ± 0.8)	(34 ± 4)	(45 ± 2)	(13 ± 1)
W	(11.9 ± 0.6)	(48 ±2)	(29 ± 1)	(11.2 ± 0.6)
FA	(7 ± 1)	(51 ± 1)	(32 ± 1)	(10 ± 1)
FA+W	(10.4 ± 0.5)	(55 ± 3)	(25 ± 1)	(9.2 ± 0.5)

[a] Reactions were carried bachwisely, no stirring, at 88-109 Bar at 280 ° C for 24 h. Ratio solid: crude oil: formic acid: water = 10:1:1:1.

By means of thermogravimetric analysis (TGA) it was possible to quantified the loss of mass as crude oil samples are heated in nitrogen atmosphere. The loss of mass was divided into four fractions according to the temperatures normally used to characterize distillation cuts in a refinery. Thus, fraction 1 corresponds to the mass lost below 175 ° C, which would be equivalent to a naphtha distilled fraction. Fraction 2 is the mass lost between 175-340 °C (equivalent to kerosene, diesel and/or medium distillates). Fraction 3 between 340-500 °C (vacuum gas oil) and fraction 4 corresponds to the residue over 500 °C. However, it is important to point out that even though TGA can be used for comparison, they can not be assumed to give the same information as simulated distillation, since in TGA not equilibrium is reached during the evaporation of the samples, and some hydrocarbon decomposition is expected to happen. Results obtained are presented in Table II.

Table II. % of mass lost, determined by TGA, for the four fractions [a]

	Fraction 1/ wt%	Fraction 2/ wt%	Fraction 3/ wt%	Fraction 4/ wt%
HAMACA	6	28	52	14
W	4	34	44	18
FA	11	26	48	15
FA+W	21	28	42	9

[a] Reaction conditions: Same as Table 1

It can be seen that when reaction is carried out in the presence of water there is an increase in fraction 2, and a decrease in fraction 3, with the other two fractions remaining at about the same level of the original crude oil. This is an indication of a rather small upgrading of the oil, due to the thermal treatment. When formic acid is used, the effect is more pronounced, thus, fraction 1 (lighter components) increases to twice the original content. However, best results are obtained when formic acid and water are used together. Fraction 4, the heaviest one, is reduced in 36 %, while fraction 1 (lighter components) increases 3.5 fold. There is also, an important reduction in fraction 3. These results are in agreement with SARA analysis, i.e., asphaltene, and resin contents (heavier fractions) are reduced when the crude oil is treated with formic acid, and formic acid together with water, and the best results are obtained for the latter.

Gas chromatographic analysis was performed on the paraffinic fractions of the Hamaca crude oil and Hamaca crude oil treated in presence of formic acid and water. The corresponding gas chromatograms are shown in Figure 1. Analysis of the total crude oil was also tried, with no success (signals could not be satisfactorily resolved). Main chromatographic peaks are in the region of C12 to C32, but the distribution is centered at around two or three lower carbon numbers for the treated oil, and more important, for the upgraded oil a new set of signals for paraffins with carbon numbers less than 12 is observed. These new signals are rather intense, and correspond to light paraffins, which is a gain in agreement with TGA and SARA results.

Sulfur and nitrogen contents for the Hamaca crude oil, before and after upgrading were also measured, and the results are presented in Table III. For the crude oil treated in presence of water a sulfur reduction of 20 %, is observed, which is due to thermal desulfurization[24]. When formic acid alone is used, the percentage of desulfurization is less (only 12%), but when formic acid and water are used together, the desulfurization is increased to 38 %, which is an important amount if we take into account that this is intended to be a downhole process.

Table III. Sulfur and Nitrogen content for original and upgraded oil. [a]

	$(S \pm 0.06)/\%$	Desulfurization/ %	$(N \pm 0.05/\%)$	Denitrogenation/ %
HAMACA	3.79	-	0.75	
W	3.02	20	0.50	33
FA	3.34	12	0.55	27
FA+W	2.36	38	0.60	20

[a] Same conditions as Table 1

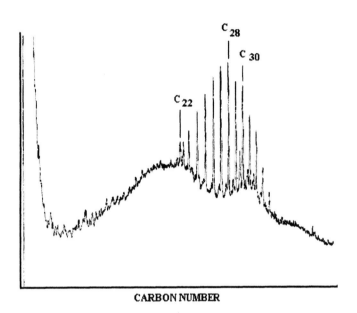

Figure 1.- GC analysis of saturate fraction of Hamaca crude oil .

As far as nitrogen content is concern, it can be observed (Table III) that an important l evel o f d enitrogenation i s o btained f or t he s ample t reated just with water. The presence of formic acid slightly reduces the amount of nitrogen removed, and the % of denitrogenation is even less when formic acid and water are used together. At the moment it is not clear why for this type of reaction a different behavior, as compared to desulfurization, is obtained.

All the results (SARA, TGA, GC and sulfur content) presented show that an important degree of upgrading is obtained when Hamaca crude oil is treated in very mild conditions, i.e. conditions that are normally used in the reservoir during steam injection, using formic acid to generate hydrogen i n t he b ulk o f reaction. The upgrading is more important when formic acid is used together with water.

Formic a cid t hermal decomposition has been proposed to proceed by two different basic mechanisms [25,26]

$$HCOOH \longrightarrow CO + H_2O \quad (1)$$
$$HCOOH \longrightarrow CO_2 + H_2 \quad (2)$$

It has been found that major products are CO_2 and H_2, for conversions between 38 to 100 %, temperatures of 320-500 °C and pressures in the range of 178-303 Bar. These conditions are similar to the ones used in this work, thus, it can be assumed that in the upgrading conditions used in this work the main decomposition pathway is the decarboxylation (reaction 2). In fact, molecular H_2 an CO_2 were observed in high proportions in the gaseous products (for example for the reaction carried out with formic acid and water the molar composition of the gas, after reaction, is: hydrogen 15.5 % v/v, nitrogen 78.2 % v/v, carbon dioxide 6.3 % v/v). On the other hand, thermodynamic calculations[26] have shown that during the hydrothermal decomposition of formic acid water acts as catalyst, by forming an intermediate between a water dimer and formic acid. It is suggested that formic acid and the water dimer are bound with two hydrogen bodings and that the water dimer acts simultaneously as a donor and acceptor of protons. This configuration greatly reduces activation energy of the decomposition process.

Results presented here show the potential of formic acid when used in conjunction with water, for generating H_2 in steam injection conditions. H_2 produced in this way is effectively used for upgrading Hamaca crude oil dispersed in sand. In the process water acts as catalysts for the thermal decomposition of formic acid, with the concomitant production of hydrogen in the bulk of the reaction system.

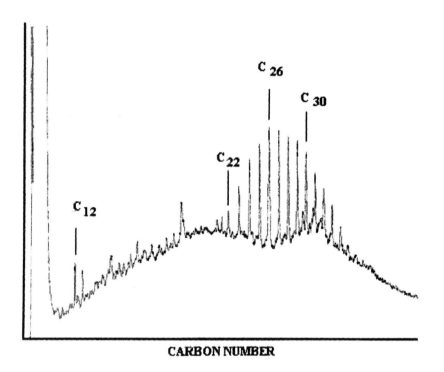

CARBON NUMBER

Figure 2.- GC analysis of saturate fraction of upgraded Hamaca crude oil, with FA+W.

Conclusions

H_2 produced by decomposition of formic acid is effectively used for upgrading of Hamaca crude oil dispersed in sand at mild conditions (280 ° C, 88-109 Bar). When formic acid was used in conjunction with water, water acts as catalysts for the thermal decomposition of formic acid, and the upgrading was more effective. Thus, for Hamaca oil upgraded in the presence of formic acid and water, a 29 % conversion of asphaltene, and 38 % of desulfurization were obtained. Also, the residue that remains after heating over 500 °C in nitrogen was reduced in 36% and the fraction below 175 ° C was increased 3.5 fold, in relation to untreated Hamaca crude oil.

Acknowledgements

The authors are grateful to FONACIT (CONICIT-CONIPET 97003783) for financial support. Also the assistance in the SARA analyses from Laboratorio de Fisicoquímica de Hidrocarburos of the UCV is gratefully acknowledged.

References.

1] Wiehe, I. A., *Am. Chem. Soc., Div. Pet. Chem. Prepr.* 2001, **46** (1), 60
2] US Patent 3 608 638, 1971 and references therein
3] US Patent 4 753 293, 1988 and references therein
4] Butler, R. M. and Mokry, I. J. SPE 25425, presented in Productions Operations Symposium, Oklahoma City, USA, March 21-23, 1993 and references therein
5] Duerksen, J. H. and Eloyan, A. M. Proceeding 6th UNITAR Int. Conf. On Heavy Crude and Tar Sands, Houston, Texas, USA, Feb 12-17, 1995, 353 and references therein
6] Henderson, J. H. and Weber, L. *J. Can. Pet. Tech.*, Oct-Dec, 1965, 206
7] Kasrale, M. and Faroug, A. SPE 18784 presented in SPE California Regional Meeting, Bakersfield, USA, April 5-7, 1989, 389
8] Shu, W. R. and Hartman, K.J. SPE *Reserv. Eng.* 1986 (September), 474
9] Monin, J. C. and Audibert, A. SPE *Reserv. Eng.* 1988 (Nov), 1243
10]Dew, J.N. Martin, W.L. US Patent 3,208,514, 1965
11]Hanrick, J.T., Rose, L. US Patent 4,050,515, 1977
12]Stine, L.O. US Patent 4,444,257, 1984
13]Ware, C.H., Rose, L.C., Allen, J.C. US Patent 4,597,441, 1986

152

14] Stapp, P. R. Report NIPER-434, Bartleswille, OK, December, 1989 and references therein

15] Weisman, J. G. and Kessler, R. V., *Appl. Catal.* 1996, **140**, 1

16] Hewgill, G.S., Kalfayan, L.J. US Patent 5,105,877, 1992

17] Brunner, G., Hoffman, R., Kunstle, K. US Patent 4,505,808, 1985

18] Roychaudhury, S., Rao, N. S., Sinha, S. K., Gupta, K. K., Sapkal, A. V., Jain, A. K., Saluja, J. S. SPE N° 37547, presented in Int. Thermal Oper. Heavy Oil Symp. Bakersfield, CA, Feb. 10-12, 1997.

19] Wichert, G. C., Okazawa, N. E., Moore, R. G., Belgrave, J. D. M. SPE N° 30299, presented in Inter. Heavy Oil Symp., Calgary, Alberta, Canada, June 19-21, 1995, 529

20] Weissman, J. G., Kessler, R. V., Sawicki, R. A., Belgrave, J. D. M., Laureshen, C. J., Metha, S. A., Moore, R. G., Ursenbach, M. G. *Ener. & Fuel* 1996, **10**, 883

21] Vallejos, C., Vasquez, T., Ovalles, C. US Patent 5,891,829, 1999

22] Vallejos C., Vasquez T. and Ovalles C. *Am. Chem. Soc., Div. Petro.l Chem. Prepr.* 2000, **45** (49), 591

23] Scott C., Alfonso H., Delgado O., Pérez-Zurita M. J., Bolívar C. and Ovalles C. *Am. Chem. Soc., Div. Petrol. Chem. Prepr.* 2000, **45** (49), 588

24] Burger, J., Sourieau, P., Combarnous, M. 'Thermal Methods of Oil Recovery', Techni, Paris, 1985, p. 81

25] Yu, J. and Savage, P. *Ind. Eng. Chem. Res.* 1998, **37**, 2

26] Wang, B., Hou, H., Gu, Y. *J. Phys. Chem. A*, 2000, **104**, 10526

Chapter 11

Thermal Fouling: Heat-Induced Deposition and Coking

John F. Schabron, Joseph F. Rovani, Jr., and A. Troy Pauli

Western Research Institute, 365 North 9th Street, Laramie, WY 82072

Heat induced deposition and pyrolytic coke formation phenomena are described using the dispersed particle solution model of petroleum residua. Below pyrolysis temperatures, heat induced deposition appears to be reversible with solvation shell energies ranging from about 0.5-1.8 kcal/mol. At pyrolysis temperatures, the protective shell surrounding asphaltene cores is irreversibly destroyed. Coking Indexes were developed to monitor proximity to coke formation during the pyrolysis induction period. Once coke formation begins, the initial amount formed is related to the original residuum free solvent volume. Coke formation was modeled using zero order kinetics as a two stage process. Activation energies for the first stage ranged from 22-38 kcal/mol, and activation energies for the second stage ranged from 54-83 kcal/mol.

Fouling and Deposition

The petroleum industry is showing an increasing interest in heavy oil and residua conversion. Such processes require heating for distillation and thermal treatment. The many sources of residua behave differently in refining processes.

The so-called opportunity resids are often purchased for their low prices. No longer does the refiner have the luxury of upgrading the same known material over time. Blends are often used, with different day-to-day compositions, resulting in unexpected fouling. Thermal fouling can occur when a material is heated for storage or pumping, or pyrolyzed in visbreaking or vacuum distillation processes. During distillation, many refiners select conservative heating profiles, which result in less than optimal distillate production. Predictability is needed to be able to better control conditions to prevent undesired deposition. In this chapter thermal fouling issues are related to the dispersed particle solution model of residua.

The Ordered Structure of Petroleum Residua

The Dispersed Particle Solution Model

Petroleum residua are highly ordered structures of polar asphaltenes dispersed in a lower polarity solvent phase. Asphaltenes are a solubility class of molecules that self associate to maintain minimum system free energy. They are isolated from a petroleum matrix by mixing the oil with an excess of a low polarity hydrocarbon solvent. The temperature of mixing is important. Higher temperatures result in significantly lower asphaltene yields than lower temperatures (1). Apparent molecular weights of asphaltenes are a function of measurement conditions such as temperature, solvent, and concentration. For example, the molecular weights of asphaltene subfractions in toluene were found to range from 1,260-23,000 g/mol when measured by vapor pressure osmometry at solution concentrations of 2-5 wt.% (2). Dilute size exclusion chromatography measurements of the same fractions at concentrations in toluene of about 0.06-0.3 wt.% yielded number average molecular weights relative to polystyrene of 523-772 g/mol. These later measurements were probably below the critical micelle concentration (CMC) of the asphaltene subfractions in toluene. Calorimetry measurements in toluene have provided a CMC value for asphaltenes near 0.38 wt.% (3).

In a whole residuum, asphaltenes interact directly in a continuum with the residuum solvent matrix. The structure is held together by intermediate polarity materials usually called resins. The components align with each other to minimize the system free energy. The ordered structure can be described by the Pal and Rhodes suspended particle solution model (4,5,6). The model considers dispersed solvated particles in a solvent matrix. The volume fraction of the core particles can be considered as the volume fraction of heptane asphaltenes ϕ_a. The volume fraction of the core is increased by a solvation shell K_S. In the model, several solvated cores bind a portion of solvent and increase the effective particle volume fraction by a term K_F. The term $K_S \cdot K_F$ is called the solvation constant K. The effective particle volume fraction ϕ_{eff} is equal to the core asphaltene volume fraction increased by the solvation terms as shown by the equation below.

$$\phi_{eff} = K\phi_a = K_F K_S \phi_a$$

For petroleum residua, typical values of K range from 3 to 6. When a residuum is heated, the value for K decreases with increasing temperature, and less trapped solvent is associated with the asphaltene structures, resulting in a flocculation of the polar core material (7,8). Below pyrolysis temperature (340 °C, 644 °F), this appears to be reversible on cooling.

Solvation constants (K) can be derived using the Pal–Rhodes equation shown below.

$$K = K_S K_F = \frac{1 - \eta_{rel}^{-0.4}}{\chi_a/1.2}$$

In the equation, χ_a is the mass fraction of heptane asphaltenes divided by an assumed density of 1.2 g/cc to yield the volume fraction of asphaltenes ϕ_a, and η_{rel} is the relative viscosity. To estimate relative viscosities, zero shear viscosities η are measured for a whole residuum and for the corresponding heptane maltenes η°. The ratio of η/η° is referred to as the relative viscosity (η_{rel}) and is diagnostic of the manner in which asphaltene particles are suspended in the maltenes. A higher relative viscosity indicates a more significant ordering.

Pauli and Branthaver (9) have demonstrated that the fraction of solvent bound in the ordered structure, α, is numerically equivalent to the parameter, p_a which is obtained from an asphaltene flocculation titration. Since $K_F = 1/(1- \alpha)$, the residua free solvent volume fraction (ϕ_{FS}) can be calculated using the equation below.

$$\phi_{FS} = 1 - K_S K_F \phi_a = 1 - 1.6 \left(\frac{1}{1 - p_a}\right)\left(\frac{\chi_a}{1.2}\right)$$

Prior work measuring K_S using a variety of experimental techniques has shown that the K_S value for unpyrolyzed residua and asphalt systems is 1.6 (9). The free solvent volume fraction ranges from 0 to 1. A small free solvent volume fraction represents a residuum that requires a highly ordered structure to maintain stability. Residua with low free solvent volume values have higher viscosities than those with relatively higher values. The ordered structure of a residuum with a relatively low free solvent volume fraction is more easily disrupted by heating than the structure of residua with relatively higher free solvent volumes (6,10).

Asphaltene Flocculation Titration

The automated asphaltene flocculation titration (AFT) is performed with the toluene soluble components of residua. Two or more toluene solutions of residua at concentrations ranging from about 5-20 wt.% are titrated with a weak solvent such as isooctane (11). The weight of residuum (W), the volume of toluene (V_s), and volume of isooctane titrant (V_t) are recorded at the flocculation point where asphaltenes just begin to precipitate for each solution. Flocculation ratio and dilution concentration terms are calculated.

$$FR = \text{Flocculation Ratio} = V_s/(V_s + V_t)$$

$$C = \text{Dilution Concentration} = W/(V_s + V_t)$$

A linear plot of FR vs. C is made, and the intercepts are determined (FR_{MAX} and C_{min}). The AFT stability parameters are defined by the equations below.

$$p_a = 1 - FR_{max} \qquad \text{Peptizability of Asphaltenes}$$

$$p_o = FR_{max} \times (1/C_{min} + 1) \quad \text{Solvent Power of Maltenes}$$

$$P = P_o/(1-P_a) = 1/C_{min} + 1 \quad \text{Overall Compatibility of Residuum}$$

Larger values of p_a and P indicate peptizable asphaltenes and an overall compatible system, respectively. A larger p_o value is subject to a mixed interpretation, however it is related to oil aromaticity. FR_{MAX} is the volume fraction of toluene in a toluene and isooctane mixture, and is a measure of the solubility parameter at which asphaltenes begin to precipitate (2). C_{min} is the ratio of residuum to titrant (isooctane) at which asphaltenes begin to precipitate, and this value can be used to calculate the solubility parameter of the whole oil.

Non-Pyrolytic Heat-Induced Fouling

Heat-induced deposition can result from the formation of asphaltene flocs when a residuum is heated above a temperature at which the intermediate polarity material no longer protects the polar asphaltene cores (7,8), but below temperatures at which pyrolysis occurs (340 °C, 644 °F). This flocculation is not the same as asphaltene precipitation due to weak solvent addition. When a residuum is heated below pyrolysis temperatures, the solvation shell is partially removed in a reversible manner and the solvation constant K decreases. On cooling, the shell is restored, although the time for restoring the original structure can vary. On heating, the exposed core material can flocculate and attach to polar metal surfaces such as surfaces in heat exchangers.

Evidence of floc particle formation was found at the bottom of aluminum weighing pans in which 5-g portions of residua were heated under argon and from which oil was decanted (6). Five residua were studied: Boscan, CA Coastal, MaxCL, Redwater B.C., and Vistar. Experiments were performed at 100 ˚C (212 °F), 175 ˚C (347 °F), and 250 ˚C (482 °F). When deposition occurred, the bottom of the pans showed a fairly uniform pattern of spots for a particular residuum. Little, if any deposition was observed at 100 ˚C (212 °F). At 175 ˚C (347 °F), the deposition patterns for the Boscan, MaxCL, and CA Coastal contained globules of oily deposits. This suggests that the deposition process is not complete at this temperature for these materials. Microscopic images (10x) of the deposition from Boscan residuum at 100 ˚C (212 °F) and 250 ˚C (482 °F) are shown in Figure 1. Little or no deposition was evident at the lower temperature, while significant deposition occurred at the higher temperature.

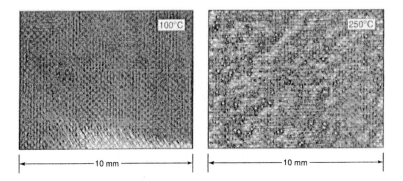

Figure 1. Heat induced deposition from Boscan residuum. (Reproduced from Reference 6, Copyright 2001 with permission from Elsevier)

At 250 °C (482 °F), deposition is significant and the deposition spot density correlates with the residua free solvent volume (Figure 2). Residua with low free solvent volumes exhibit higher heat induced deposition tendencies than residua with relatively higher free solvent volumes.

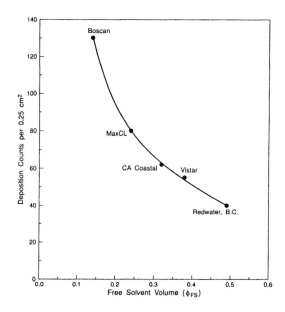

Figure 2. Deposition density correlation with residua free solvent volume. (Reproduced from Reference 6, Copyright 2001 with permission from Elsevier)

Storm et al. (8) suggest that the polar core particles are fully unprotected at 200 °C (392 °F) for the system they studied. On heating, the intermediate polarity material surrounding the polar cores appears to be driven into solution, exposing the polar cores. These cores flocculate and some of the polar flocs interact with the polar metal surface. When the oil is decanted, some of the flocs remain bound to the metal. The flocs cannot return to solution, since the oil has been decanted off. The oily spots observed on cooling are possibly a combination of the polar flocs and some remaining oil following the pouring. The deposition is possibly similar to the residue that appears on steel thimbles heated to 177 °C (350 °F) for six hours for fuel oils tested by ASTM Method D 1661, Thermal Stability of U.S. Navy Special Fuel Oil (12).

Deposition experiments were performed with three portions of whole Lloydminster residuum at 250°C (482 °F) with and without 0.2 wt.% of commercial anti fouling additives (13). The materials with the additives showed some spots, but overall smoother deposition profiles were evident when compared to the original material. The additives appear to suppress agglomerization and cause a smoother dispersion of the deposition pattern than when they are not present.

Solvation shell energies for the dissipation of the protective layer around the polar asphaltene cores were estimated from the slopes of ln K vs. 1/T plots using relative viscosity measurements at several temperatures for Boscan, CA Coastal, MaxCL, Redwater B.C., and Vistar residua. The energies ranged from about 500-1,600 cal/mol, which is similar to the energy of about 1,800 cal/mol for an Arabian residuum (8). These results suggest that to minimize fouling when residua are heated for pumping or storage, the lowest possible temperature should be used to keep the polar asphaltene cores as solvated as possible.

Metals and Porphyrins Determinations for Deposition Spots

To explore the potential asphaltene nature of the deposition spots, heated pouring experiments were performed with Boscan residuum. The material remaining at the bottom of the weighing pan after pouring at 250 ˚C (482 °F) and 300 ˚C (572 °F) and the original residuum were dissolved in methylene chloride. UV-visible spectra were obtained to evaluate the porphyrin, or Soret peak at 410 nm. In addition, nickel and vanadium were determined for these materials by inductively coupled plasma (ICP) spectroscopy. The results of these analyses are provided in Table I. The deposits are enriched in porphyrins, nickel and vanadium relative to the original residuum. The concentrations of metals and porphyrins are greater for the 300 ˚C (572 °F) deposition experiment than for the 250 ˚C (482 °F) experiment, which is consistent with the presence of asphaltene material.

Table I. Porphyrins and Metals for Boscan Deposition Experiments, mg/kg

Material	Porphyrin (Ni+V)	Ni	V	(Ni + V)
Original Residuum	490	126	1300	1430
Pour Pan Deposits 250 ˚C	520	155	1430	1580
Pour Pan Deposits 300 ˚C	590	203	1640	1840

SOURCE: Reproduced from Reference 6, Copyright 2001 with permission from Elsevier.

Deposition on Teflon and Stainless Steel

Deposition experiments were performed using Boscan residuum at 250 °C (482 °F) under argon with Teflon and 304 stainless steel discs placed at the bottom of pouring pans (6). This was done to explore the spot patterns with a nonpolar surface and on a metal commonly used in refineries. On the Teflon, only a few droplets of oil were evident, with the surface being mainly clear. The aluminum pan surface around the Teflon disc showed typical spot patterns observed in experiments without the disc. The spot pattern the stainless steel disc was similar to that observed on the aluminum. The results suggest that deposition problems might be minimized on critical heated surfaces using non-polar surface materials such as silanized or deactivated stainless steel.

New Coking Indexes

Refinery processes that convert heavy oils and residua to lighter distillate fuels require heating for distillation, hydrogen addition, or carbon rejection (coking). The efficiency of conversion is limited by the formation of insoluble carbon-rich coke deposits (14,15). A common problem in the refining industry is to ascertain how close a pyrolysis system is to forming coke. This has resulted in undesired coking events in distillation units, causing in significant downtime and economic loss. To minimize the risk, conservative heating profiles are often used, which results in less than optimal distillate yield, and less profitability.

When a residuum is heated above the temperature at which pyrolysis occurs (340 °C, 650 °F), there typically is an induction period ranging from a few seconds to over an hour after which coke formation begins (16,17,18). As pyrolysis progresses, the intermediate polarity material decreases during the induction period. Pyrolysis reactions involve the cleavage of carbon bonds with carbon, hydrogen, and heteroatoms resulting in the formation of free radicals that continue scission reactions or condense into carbon-rich material (19,20). The apparent molecular weight of asphaltenes in toluene measured by VPO decreases, and they become more polar (2). As the protective shell around the asphaltene cores is destroyed, they agglomerate within the whole oil matrix, and the particle size of the agglomerated complexes increases. For example, asphaltene-type particles in unpyrolyzed Lloydminster residuum were visualized by atomic force microscopy to be about 0.5 micron in diameter (21). At a pyrolysis time of 15 minutes at 400 °C (752 °F), the particles increased to a diameter pf about 2 microns. At 30 minutes, they were about 5 microns. At 60 minutes, coke formation had begun. New Coking Index concepts were developed to measure how close a residuum is to coke formation when it is heated above pyrolysis temperatures (2,22). These involve relatively simple experimental approaches as described below.

Ratio of p_a/C_{min} from Flocculation Titration

One Coking Index is defined as the ratio of the AFT parameters, p_a/C_{min} (2,22). This is based on the considerations that as pyrolysis progresses, p_a decreases as the polarity of the asphaltenes increases, and C_{min} increases as the overall stability of the oil matrix decreases. This ratio can be used as a Coking Index to measure the proximity to coke formation during pyrolysis. The initial ratio value for a stable residuum can range from 1-2 or higher. Coke formation is plotted against the p_a/C_{min} Coking Index in Figure 3 for a variety of residua. The coke formation threshold value is about 0.2, below which coke forms. Once coke is present, the product oil consists of an unstable multi-phase liquid system, and it is difficult to obtain reliable AFT data from the material. The residuum with the p_a/C_{min} value of 1.5 in Figure 3 represents a material for which distillation was stopped early. By continuing distillation to a Coking Index value of 0.6-0.8, for example, additional distillate could be obtained from this material without undesired coke formation.

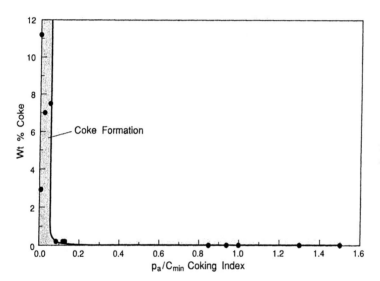

Figure 3. Coke formation vs. p_a/C_{min} Coking Index. (Reproduced from Reference 22, Copyright 2001 with permission from Elsevier)

Cyclohexane Soluble Asphaltenes

In an unpyrolyzed residuum, as much as 50 wt.% of the heptane asphaltenes are soluble in cyclohexane, which has a solubility parameter midway between toluene, in which the asphaltenes are fully soluble, and heptane (2,22). As pyrolysis progresses, less of the heptane asphaltenes dissolve in cyclohexane due to the destruction of the intermediate polarity resins material. Coke begins to form when the depletion of the resins disables the ability of the asphaltene/resin complexes to self-adjust their apparent molecular weights to closely match the solubility parameter of the matrix. The bimodal system formed has been visualized as two immiscible liquid phases by atomic force microscopy and magnetic resonance imaging (21,23,24).

The asphaltene solubility Coking Index is obtained by determining the weight percent the weight percent heptane asphaltenes (X) and the weight percent of cyclohexane soluble material in heptane asphaltenes (Y). The ratio Y/X is calculated. The initial value for a stable residuum can range from 2 - 4 or higher. As pyrolysis proceeds, the amount of cyclohexane soluble asphaltenes decreases and the Y/X value decreases to a threshold value of about 1.0, below which coke begins to form (Figure 4). Unlike the AFT titration, the Y/X analysis can be reliable when performed after coke formation begins. The Coking Indexes do not predict the amount of coke formed; however, the initial amount of coke formed with pyrolysis at a particular residence is influenced by the unpyrolyzed residua free solvent volume, as described in a subsequent section.

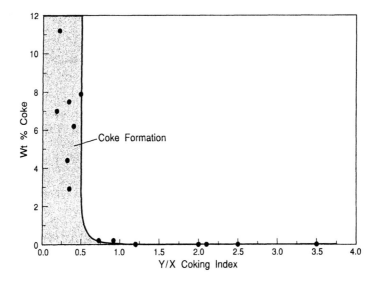

Figure 4. Coke formation vs. Y/X Coking Index. (Reproduced from Reference 22, Copyright 2001 with permission from Elsevier)

Characterization of Pyrolysis Product Oils and Asphaltenes

Boscan residuum was pyrolyzed for 25 and 55 minute residence times at 400 °C (752 °F). The amount of coke and volatiles and the WRI Coking indexes for the starting and product oils are listed in Table II. The 25 minute product oil did not contain any coke, and the 55 minute product oil contained on average 6.4 wt. % coke. Data for the original and pyrolyzed product oils are provided in Table III. The H/C ratios of the original unpyrolyzed oil and the heptane asphaltenes are 1.5 and 1.2, respectively. As pyrolysis progresses, the H/C ratio of the whole product oil is nearly constant, at 1.4-1.5. The H/C ratio for the asphaltenes decreases significantly with increasing residence time. From the elemental total values, it is apparent that the asphaltenes probably contain about 2 wt.% oxygen by difference. The nitrogen content of asphaltenes and coke is 2-5 times higher than in the corresponding maltenes. The sulfur content of asphaltenes and coke is only slightly higher than in the corresponding maltenes, about 1-2 wt. %. The elemental composition of cyclohexane soluble and insoluble asphaltenes is similar for a given residence time. Pyrolysis destroys the porphyrin structure that absorbs light at 410 nm., and the porphyrin content of the maltenes is lowered during pyrolysis. The porphyrins content of the original and pyrolyzed asphaltenes is about the same. The vanadium content of any of the oils or subfractions is about ten times greater than the nickel content of the particular oil or subfraction. Metals (Ni,V) are transferred from the maltene phases into the asphaltenes with pyrolysis. The metals content of the cyclohexane soluble and insoluble asphaltenes is similar for the unpyrolyzed oil. For the 25 minute pyrolysis oil, the metals content of the cyclohexane insoluble asphaltenes is about 64% higher than the cyclohexane soluble asphaltene material. The metals are concentrated into the coke once coke begins to form (25,26).

Table II. Coking Index Data for Original and Pyrolyzed Boscan Residuum

Boscan Residuu	Wt. % Asphaltenes				Wt.%		Coking Indexes	
	X: n-C7	Y: CyC6 Sol.	p_a	C_{min}	Coke	Volatiles	Y/X	p_a/C_{min}
0 min.	18.4	37.4	0.662	0.441	<0.05	-	2.04	1.51
25 min.	18.4	16.4	0.463	0.919	<0.05	1.4	0.89	0.49
55 min.	13.5	4.9	nd	nd	6.4	2.2	0.36	nd

SOURCE: Reproduced from Reference 26.

Table III. Elemental and Metals Analysis Data for Original and Pyrolyzed (400 °C, 752 °F) Boscan Residuum and Fractions

Boscan Residuum	Weight Fraction	Weight Percent					H/C Ratio	mg/kg		
		C	H	N	S	Total		Ni	V	Porph.
0 min.	1.000	83.1	10.3	0.8	5.5	99.7	1.5	130	1300	80
Maltenes	0.816	83.2	10.6	0.5	5.0	99.3	1.5	59	530	350
Asphaltenes	0.184	81.6	7.9	1.8	6.6	97.9	1.2	450	4700	1100
CyC6 Sol.	*0.360*	81.4	8.0	1.7	-	-	1.2	390	4000	1100
CyC6 Insol.	*0.640*	81.7	7.8	1.9	6.6	98.0	1.2	470	4800	1000
Balance to Whole	1.000	82.9	10.1	0.7	5.3	-	1.5	130	1300	490
25 min.	1.000	83.6	9.7	0.5	5.0	98.8	1.4	140	1300	310
Maltenes	0.811	83.9	10.8	0.4	4.5	99.6	1.5	3℃	300	180
Asphaltenes	0.189	82.3	6.7	2.3	6.4	97.7	0.98	580	6100	1000
CyC6 Sol.	*0.182*	82.5	7.2	1.9	-	-	1.0	370	4100	1100
CyC6 Insol.	*0.818*	82.3	6.5	2.4	6.5	97.7	0.95	610	6700	1000
Balance to Whole	1.000	83.6	10.0	0.8	4.8	-	1.4	140	1400	330
55 min.	1.000	84.3	9.5	0.8	4.8	99.4	1.4	140	1400	180
Maltenes	0.801	83.8	10.6	0.4	4.1	98.9	1.5	21	120	36
Asphaltenes	0.135	83.2	6.2	2.2	5.9	97.5	0.89	450	5000	1000
CyC6 Sol.	*0.048*	-	-	-	-	-	-	-	-	-
CyC6 Insol.	*0.952*	-	-	-	-	-	-	-	-	-
Coke	0.064	83.2	5.4	2.6	6.2	97.4	0.78	770	7700	-
Balance to Whole	1.000	83.6	9.7	0.8	4.5	-	1.4	130	1300	160

SOURCE: Derived from Reference 26.

Predicting Amount of Initial Coke Formation

Initial coke formation predictability maps were developed using four residua, Boscan, MaxCL2 (Conoco), Lloydminster, and Redwater B.C. All four residua had atomic H/C ratios of 1.4 within experimental error. The asphaltene contents of Boscan, MaxCL2, and Lloydminster were virtually identical, ranging from 16.9 to 17.6 wt %. The asphaltene content of the Redwater, B.C. is significantly lower at 8.9 wt %. The only measurements required for the calculation of the free solvent volume fraction ϕ_{FS} is the weight percent heptane asphaltenes and the AFT p_a value. The free solvent volume fractions for the oils were significantly different: Boscan 0.22, MaxCL2 0.32, Lloydminster 0.40, and Redwater B.C. 0.66

Three-dimensional coke formation maps were developed at 400 ˚C (752 ˚F), 450 ˚C (842 ˚F), and 500 ˚C (932 ˚F) (10). The maps are comprised of plots of free solvent volume (x), amount of coke formed (y), and pyrolysis residence time (z). The initial amount of coke formed correlates with residence time and free solvent volume at 400 °C (752 °F) and 450 °C (842 °F). At 500 °C (932 °F), coke formation is very rapid. The maps at 400 °C (752 °F) and 450 °C (842 °F) are shown in Figure 5.

The structure of residua with low free solvent volumes is more easily destroyed during pyrolysis than the residua with relatively higher free solvent volumes, and thus they produce more coke initially than residua with relatively higher free solvent volumes. When carried to completion, the amount of coke formed is related to the atomic H/C ratio. The results provide a new tool for ranking residua and predicting initial coke make tendencies.

Coke Formation Activation Energies

Coke formation was studied using four residua, Boscan, MaxCL2 (Conoco), Lloydminster, and Redwater B.C. (10). The coke formation plots at 400 ˚C (752 ˚F), 450 ˚C (842 ˚F), and 500 ˚C (932 ˚F) for Boscan residuum is provided in Figure 6. Linear regression was used to determine ths slopes of the 400 ˚C (752 ˚F) and 450 ˚C (842 ˚F) lines. The approach to determine the slope of the at 500 ˚C (932 ˚F) line is described in Reference 10. Activation energies were calculated for the four residua using the slopes. The plots for the four residua are all qualitatively similar. These mostly are linear in nature, indicating apparent zero-order (decomposition) kinetics.

A coke formation activation energy of 42 kcal/mol was calculated for Athabaska bitumen assuming first-order kinetics (17). First-order kinetics were assumed for coke formation from Athabaska bitumen, however the data show linear coke formation with time at 400 °C (752 °F), suggesting that this process proceeds with zero-order kinetics (20). In a study of the kinetics of pyrolysis of Cold Lake residuum, data showed first-order kinetics for coke formation from asphaltenes (18). Zero-order kinetics were evident from data for coke formation from maltenes pyrolysis. Coke formation data for whole Cold Lake residuum suggested first-order kinetics. Coke formation most likely involves a complex suite of reactions that on the whole fall somewhere between apparent zero-order and first-order kinetics when viewed in terms of the appearance of coke.

166

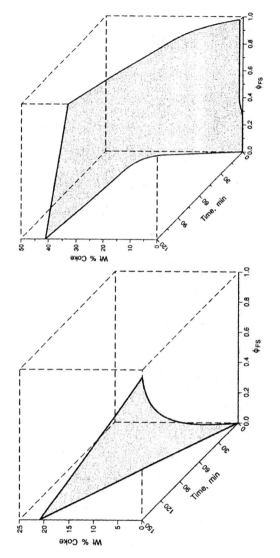

Figure 5. Coke formation map at 400 °C (752 °F) (left) and 450 °C (842 °F) (right). (Reproduced from Reference 10, Copyright 2002 with permission from Elsevier)

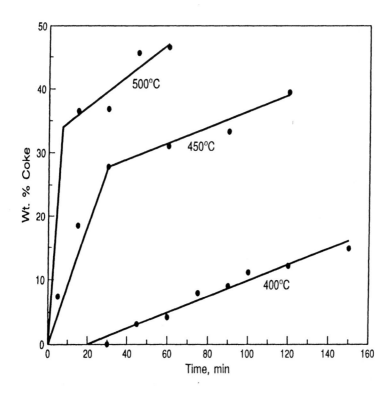

Figure 6. Coke formation for Boscan Residuum. (Reproduced from Reference 10, Copyright 2002 with permission from Elsevier)

Two different coke formation mechanisms or processes are evident from the change in slope in the 450 °C (842 °F) line. The first is represented by the line at 400 °C (752 °F) and the lower line at 450 °C (842 °F). The second is represented by the upper line at 450 °C (842 °F). The coke formation process at 500 °C (932 °F) is very rapid up to the limiting amount of coke (determined by the H/C ratio), which is evidenced by the break in the slope of the upper portion of the 500 °C (932 °F) plot. The change in slope at the top of the 500 °C (932 °F) line is most likely due to the upper limit of coke formation having been reached. The activation energies for the initial first stage coke formation reactions are estimated to range from 22-38 kcal/mol. The activation energies for the secondary coke formation reactions range from 54-83 kcal/mol. Some of the secondary coke formation reactions could also involve wall effects in the reactor tubes. The initial coke formation represents 58.1–63.6% of the final coke yield for the four residua. This is consistent with observations with Athabaska bitumen in which the first 40–50% of conversion reactions involve the cleavage of side chains followed by condensation aromatic carbon radical coke-producing reactions (27). Activation energies were derived for cracking and coke formation reactions of 41 and 64 kcal/mol, respectively; assuming first-order kinetics for a Belaym vacuum residuum (20). The above results suggest that heat and time are not fully interchangeable variables in thermal conversion. To maximize conversion yield, the feed should be kept at a temperature low enough to maximize the bond cleavage reactions and minimize the condensation reactions.

Table IV. Estimated Activation Energies (cal/mol) of Coke Formation

Residuum	Primary Coke Formation	Wt % Primary (450 °C)	Secondary Coke Formation	Wt % Total (500 °C)
Boscan	38,000	27.0	83,000	58.1
MaxCL2	30,000	28.5	78,000	63.6
Lloydminster	30,000	24.5	66,000	62.5
Redwater B.C.	22,000	22.5	54,000	59.9

SOURCE: Reproduced from Reference 10, Copyright 2002 with permission from Elsevier.

There also appears to be a correlation between the activation energies for both the primary and secondary coke formation reactions and the residuum free solvent volume (10). A residuum with a lower free solvent volume that is indicative of the presence of a more highly ordered system has higher activation energies than a residuum with a higher free solvent volume.

The physical difference between the primary and secondary coke is that the primary, or initial coke made is mostly in the form of a suspension in the liquid product oil. Once the secondary coke begins to form at 450 °C (842 °F), the coke takes the form of a hard, solid, cross-linked mass that adheres to the reactor tube walls. At 500 °C (932 °F), all of the coke is of the latter variety.

Acknowledgments

Funding for this study was provided by the U.S. Department of Energy under Cooperative Agreement DE-FC26-98FT40322, and by ChevronTexaco, ConocoPhillips, ExxonMobil Research and Engineering, and GE Betz. The authors would like to acknowledge Tony Munari for preparing the figures.

Disclaimer

References

1. Andersen, S.I., Keul, A., Stenby, E. *Petroleum Science and Technology*, **1997**, 15 (7&8), 611–645.
2. Schabron, J.F., Pauli, A.T., Rovani, J.F. Jr. *Fuel*, **2001**, 80 (4), 529–537.
3. Andersen, S.I. Birdi, K.S. *Journal of Colloid and Interface Science*, **1991**, 142, 497-502.

4. Pal R., Rhodes, E. *Journal of Rheology*, **1989**, 33 (7), 1021–1045.
5. Pauli, A.T., Branthaver, J.F. *Prepr. Pap. Am. Chem. Soc. Div. Pet. Chem.*, **1999**, 44(2), 190-193.
6. Schabron, J.F., Pauli, A.T., Rovani, J.F. Jr. *Fuel*, **2001**, 80 (7) 919–928.
7. Storm, D.A., Barresi,R.J. Sheu, E.Y. *Energy and Fuels*, **1995**, 9, 168-176.
8. Storm, D.A., Barresi,R.J., Sheu, E.Y., *Fuel Science and Technology International*, **1996**, 14(1&2), 243-260.
9. Pauli, A.T., Branthaver, J.F. *Petroleum Science and Technology*, **1998**, 16 (9&10), 1125–1147.
10. Schabron, J.F., Pauli, A.T., Rovani, J.F.Jr. *Fuel*, 81, **2002**, (17) 2227-2240.
11. Heithaus, J.J. *Journal of the Institute of Petroleum*, **1962**,48 (458), 45–53.
12. ASTM D1661, Standard Test Method for Thermal Stability of U.S. Navy Special Fuel Oil (Discontinued in 1990), *ASTM International Volume 5.02*, 1988.
13. Schabron, J.F., Rovani, J.F. Jr.*Prepr. Pap. Am. Chem. Soc. Div. Fuel Chem.*, **2003**, 48 (1) 96.
14. Parker, R.J., McFarlane, R.A., *Energy and Fuels*, **2000**, 14: 11–13.
15. Brons, G., Wiehe, I.A. *Energy and Fuels,* **2000**, 14: 2–5.
16. Magaril, R.Z., Aksenova, E.I. *International Chemical Engineering*, **1968**, 8 (4), 727.
17. Phillips, C.R., Haidar, N.I., Poon, Y.C. *Fuel*, **1985**, 64(5), 678-691.
18. Wiehe, I.A. *Ind. Eng. Chem. Res.*, **1993**, 32 (11), 2447–2454.
19. Singh, I.D., Kothiyal, V. Ramaswamy, V., Krishna, R. *Fuel*, **1990**, 69 (3), 289–292.
20. Del Bianco, A., Panariti, N., Anelli, M., Carniti, P. *Fuel*, 72, **1993**, 2 (1), 75–80.
21. Schabron, J.F., Miknis, F.P., Netzel, D.A., Pauli, A.T., Rovani,J.F. Jr. *Prepr. Pap. Am. Chem. Soc. Div. Pet. Chem.*, **2002**, 47 (1), 17-21.
22. Schabron, J.F., Pauli, A.T., Rovani, J.F. Jr., Miknis, F.P. *Fuel*, **2001**, 80 (11), 1435–1446.
23. Taylor, S.E. *Fuel*, **1998**, 77 (8), 821–828.
24. Shenghua L., Chenguang, L.,Guohe, Q., Wenjie, L., Yajie, Z. *Petroleum Science and Technology*, **1999**, 17 (7&8), 693–709.
25. Reynolds, J.G., Biggs, W.R.*Prepr. Pap. Am. Chem. Soc. Div. Pet. Chem.*, **1985**, 30 (4), 679-686.
26. Schabron, J.F., Rovani, J.F. Jr., Miknis, F.P. and Turner, T.F. Residua Upgrading Efficiency Improvement Models: WRI Coking Indexes, *WRI Topical Report to USDOE under contract DE-FC26-98FT40322*, 2003.
27. Sanford, E.C. *Prepr. Pap. Am. Chem. Soc. Div. Pet. Chem.*, **1993**, 38 (2), 413–416.

Chapter 12

Coking Reactivities of Petroleum Asphaltenes on Thermal Cracking

Ryuzo Tanaka[1], Shinya Sato[2], Toshimasa Takanohashi[2], and Satoru Sugita[3]

[1]Central Research Laboratories, Idemitsu Kosan Company Ltd., 1280 Kamiizumi, Sodegaura 299–0293, Japan
[2]Institute for Energy Utilization, National Institute of Advanced Industrial Science and Technology, 16–1, Onogawa, Tsukuba 305–8569, Japan
[3]Technical Development Group, Kobe Steel, Ltd., 2–3–1, Shinhama, Araicho, Takasago 676–8670, Japan

The autoclave experiments were performed to clarify the thermal cracking reactivities of three petroleum asphaltenes (Maya, Khafji, and Iranian Light). The influence to the reactivities of solvents (1-methylnaphthalene, quinoline, and decalin) were investigated in the point of view of asphaltene aggregates relaxation and hydrogen donation. Asphaltenes were cracked with/without the solvents at 440 °C under nitrogen pressure of 8 MPa in the 30 mL autoclave. Quinoline seems to have a function of the asphaltne aggregates relaxation, while decalin seems to have both of the functions that the asphaltene aggregates relaxation (but less than quinoline) and hydrogen donation. For Khafji and Iranian Light asphaltenes, decalin is the most effective solvent for suppression of coking on the thermal cracking. On the other hand, quinoline/decalin mixture is the best for Maya asphaltene. Maya asphaltene may need more relaxation of aggregates to be donated hydrogen from decalin effectively than others, because it aggregates tighter than the other asphaltenes.

Introduction

Petroleum asphaltene is the heaviest portion of the oil fractions and may cause coking or plugging troubles in the refinery or in the transportation lines. Asphaltene is defined operationally by its solvent solubility, and it is an extremely complex organic mixture.[1]

Molecular structure of asphaltene is considered to rule the chemical reactivity of it, and many studies were done along this line. It is also supposed that the cracking and coke formation reactivities of asphaltenes are dominated not only by their molecular structure but also by their aggregation tendency.[2] In other word, no-covalent bond between asphaltene molecules may play a certain role in those reactions. The interactions contributing to the asphaltene aggregation have not yet completely clarified, because of the complexity both of the asphaltene structures and aggregation phenomena. Although, there must be some rigorous relationships among the molecular structure, the aggregating tendency, and the coking reactivity of asphaltenes.

Recently, some fundamental studies about molecular structure and aggregation phenomena of three petroleum asphaltenes (Maya, Khafji, and Iranian Light) with ^{13}C-NMR,[3] molecular dynamics simulation,[4] small angle neutron scattering (SANS),[5] and small angle X-ray scattering (SAXS)[6] were carried out. From those studies, some asphaltene aggregates sustain up to around 350–400 °C, at the temperature some chemical reaction start to occur.

As a mean to relax the asphaltene aggregates, usage of a solvent may be effective. Some solvents have abilities not only of dissolving asphaltenes but also of the hydrogen donation to that. Many studies about those solvent abilities are reported in the fields of coal liquefaction technologies.[7-13]

In this study, coking reactivities of three petroleum asphaltenes on thermal cracking were investigated with some set of autoclave experiments. The main concerns of the paper are whether the relaxation of the asphaltene aggregation has the influence to coking reactivities or not, and how to handle the VR or asphaltenes with heat and the solvent treatment to suppress the coking.

Experimental Section

Sample preparation and analyses. The residua (> 500 °C) was obtained by vacuum distillation of three crude oils. Asphaltenes were isolated by addition of a 20:1 excess of n-heptane to each of the residue at 25 °C. The suspension was stirred 1 h at 100 °C in the autoclave. After cooling down and leaving at 25 °C overnight, the suspension was filtrated. The precipitate was washed with n-heptane twice and dried. The yields of asphaltenes (precipitates) of Maya, Khafji, and Iranian Light are 24.9, 14.2, and 6.3wt%, respectively. In the separation procedure, abstraction of toluene insoluble fractions were omitted to accommodate for the mass separation of asphaltenes available in many autoclave testing, based on the check that the asphaltene content was consistent with a standard method (ASTM D4124). Table 1 shows properties of the maltenes and asphaltenes. Carbon and hydrogen contents were measured using a CHN600 (LECO), sulfur, nitrogen, and oxygen contents were measured using an AQS-6W sulfur tester (Tanaka Scientific Instrument), an ANTEK7000 (Antek), and a varioEL III (Elementar), respectively. Metals were determined by the induced coupled plasma (ICP) method using a SPS1500VR Plasma Spectrometer (Seiko Instruments). Densities were measured in conformity with JIS K 7112 using a DMA45 (Paar). Molecular weights were measured using an Automatic Molecular Weight Apparatus (Rigosha). Carbon disulfide was used as solvent for the VPO measurements.

Asphaltene cracking and separation. All reaction experiments were carried out batch wise in a 30 mL micro reactor. In a typical experiment, 2 g of asphaltene and 8 g of solvent were loaded into the reactor. A stainless ball was also loaded for stirring. The reactor was then pressurized with nitrogen of 5 MPa at room temperature and was immersed in a preheated tin bath at 300 °C. In the case with the relaxation of asphaltene aggregates, tin bath temperature was hold at the temperature for 60 min to physically agitate the sample. Then the temperature of tin bath was rose with the heating rate of 2 °C/min until the temperature reached 440 °C, the nitrogen pressure becomes ca. 8 MPa at the temperature. The reactor was oscillated all the time in tin bath. After the reaction was completed, the reactor was cooled in air. In the case without the relaxation of asphaltene aggregates, the temperature of tin bath start rising just after the reactor was immersed in it. 1-Methylnaphthalene (1MN), quinoline (Qui), decalin (Dec), and the mixture of quinoline and decalin in the ratio of 50/50 (Q+D) were used as the solvent.

Gas yields were calculated from the weight decrease of the reactor with releasing gas pressure after reaction. Liquid product and coke were washed out of the reactor with tetrahydrofuran (THF), then THF was evaporated. The product were separated with n-heptane and toluene into three fractions, heptane soluble (HS), heptane insoluble and toluene soluble (HI-TS), and toluene insoluble (TI).

Results and Discussion

Asphaltene properties. As shown in table 1, Maya asphaltene (As-MY) is the heaviest, having the highest density, and contains the most amount of metals among the three asphaltenes. Khafji asphaltene (As-KF) is medium heavy and contains the least amount of metals. H/C atomic ratio of As-KF is the highest, and this could be interpreted as the lowest aromaticity. Iranian Light asphaltene (As-IL) is the lightest, having the lowest density and smallest molecular weight, but contains much nitrogen and metals. It also has the lowest H/C atomic ratio, meaning the highest aromaticity. These properties are considered to affect aggregation phenomena of asphaltene molecules.

Table 1. Properties of asphaltenes.

source of VR	Maya		Khafji		Iranian Light	
abbreviation	MY		KF		IL	
fraction	maltene	asphaltene	maltene	asphaltene	maltene	asphaltene
yields in VR, wt%	75.1	24.9	85.8	14.2	93.7	6.3
elemental, wt %						
carbon	83.4	82.0	83.4	82.2	84.5	83.2
hydrogen	10.4	7.5	10.6	7.6	10.9	6.8
sulfur	4.6	7.1	4.5	7.6	2.9	5.9
nitrogen	0.4	1.3	0.3	0.9	0.5	1.4
oxygen	0.5	1.2	0.4	1.1	0.4	1.5
H/C	1.50	1.10	1.53	1.11	1.55	0.98
metals, wtppm						
Ni	46.3	390	19.4	200	44.7	390
V	233	1800	62.9	550	128	1200
Mn (VPO)	720	4000	610	4000	530	2400
density, g/cm^3	1.0367	1.1767	1.0148	1.1683	1.0059	1.1669

Reactivities of vacuum residua. The yields of the vacuum residue thermal cracking without solvent are shown in Table 2. In this paper, TI is called coke and the discussion is based on the view that the reaction with the lower coke yields is the more favorable. The order of the coke yields versus feed vacuum residua is MY > KF > IL. It is the same order of the coke yield vs. feed asphaltene. The coke is supposed to be converted mainly from asphaltenes in vacuum residues. (As shown in later, the coke yields from asphaltenes are higher than that versus asphaltene contents in vacuum residua.) It is supposed generally that the factors which dominant the coke yields are the amount of asphaltenes in the feed, quality of the asphaltenes, quality of the maltenes, and the compatibility between asphaltenes and maltenes.

Table 2. Product yields of vacuum residues thermal cracking. (440 °C, 1 h, N₂: 8 MPa, unit: wt%)

	VR-MY	VR-KF	VR-IL
Gas	3.3	2.3	1.7
HS	66.5	75.1	85.4
HI-TS	16.7	18.2	12.4
TI	13.6	4.4	0.4
TI vs. As	54.5	31.3	6.9

Reactivities of asphaltenes. Figure 1 shows the yields of asphaltene cracking without solvents. The coke yields converted from asphaltenes are around 50% and more than the coke yields from vacuum residue cracking in vs. feed asphaltenes. It is supposed to indicate that the maltenes have more effects than just only diluting the asphaltenes as solvent. In the case of the asphaltenes without solvents or maltenes, physical agitation for 1 h at 300 °C is no use to reduce the coke yields. The order of the coke yields from three asphaltenes is MY > IL > KF in both the case with and without the agitation.

The coke yields in the case with agitation and some asphaltene properties, densities, C/H, and metal contents are shown in Figure 2. Among the three properties, the metal content corresponds best with coke yields. The recent thermal analysis study by Zhang *et al.* revealed that metals in organic metal compounds promote the coking reaction of the organic portion in the compound

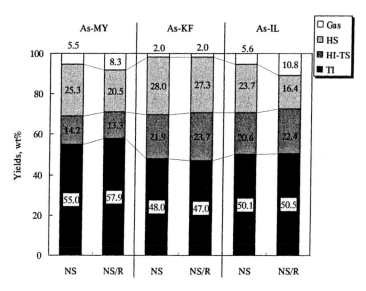

Figure 1. Yields of asphaltene cracking without solvent. (NS: no solvent, /R: relaxation of asphaltene aggregates by agitation for one hour at 300 °C)

molecues, although the metals do not influence the coking reactions of other molecules. From the result, the coke yields produced by the influence of organic metals supposed to be less than 1 wt%. Therefore, high metal content asphaltenes may have higher coking tendency owing to the carbon structure rather than metal content.

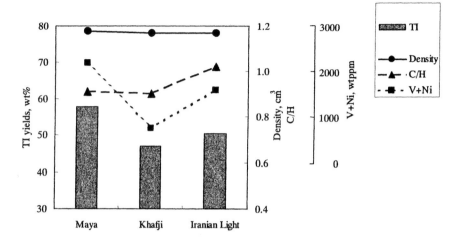

Figure 2. Coke yields and asphaltene densities, C/H, and metal contents.

The difference of the order in the coke yields from three asphaltenes between VR and asphaltene shows the influence of maltene. To clarify the effects of maltenes as a solvent, some model solvents were examined. Figure 3 shows the yields of the asphaltene cracking with quinoline (Qui). The coke yields of the asphaltene cracking decrease about 10 wt% than the case without solvent, and the order of coke yields from three asphaltenes is the same with the case of without solvent (MY > IL > KF). With quinoline, the relaxation of asphaltene aggregates by stirring at 300 °C for one hour decrease the coke yields with ca. 5 wt% of all asphaltene cracking reactions. It indicates that the physical agitation is not effective to relax the asphaltene aggregtes without solvents, but is effective with some appropreate solvents.

Reactivities of asphaltene with solvents. To understand the function of the solvents more clearly, 1-methylenaphthalerene (1MN) and decalin (Dec) and mixture of quinoline and decalin (Q+D) were used likewise quinoline. After here all the data were gained from the experiments with physical agitation for one hour at 300 °C. Figure 4 shows the yields from As-MY with or without solvent (NS). The order of coke yields with solvents is 1MN > Qui > Dec > Q+D. With quinoline and decalin mixture, the coke yield is less than with quinoline or decalin solely. It infers that there exists the synergetic effect with the two solvents for As-MY.

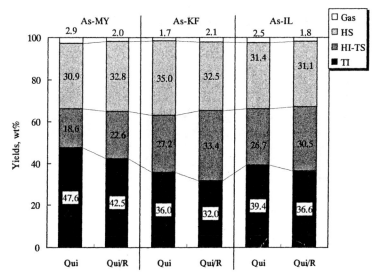

Figure 3. Yields of asphaltene cracking with quinoline.

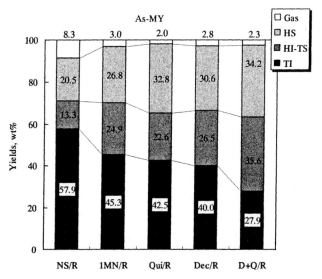

Figure 4. Yields of As-MY cracking without and with some solvents.

Figure 5 shows the yields from As-KF. The order of TI yields with three solvents are 1MN > Qui > Q+D >> Dec. Decalin shows the largest effect to suppress the coking and the TI yields of Q + D is almost average of Qui and Dec. It is reported so far that decalin is a hydrogen donor[14-16] or not at all[17,18] depending on the reaction feeds and conditions. In this case, decalin seems to show the hydrogen donor ability. Indeed, the recovered solvent after reaction experiments contained the dehydrogenated decalin compounds (the content of tetralin and naphthalene are 1.1 and 0.2 mol% of feed Decalin, respectively).

The yields of As-IL cracking have higher TI yields with each solvent than As-KF and show the same tendency depending on the solvent with As-KF as shown in Figure 6. On the other hand, As-MY shows the different TI yields order with the solvents. The yields of As-MY craking is shown in Figure 6.

Solvent synergy effect. Comparing the order of the coke yields of each asphaltenes with solvents, it may be interpreted that both aggregates of As-KF and As-IL breake up even in decalin at around reaction temperature and hydrogen is easily donated to the asphaltenes, but As-MY remains the aggregates in the decalin up to at around reaction temperature. In other word, quinoline assists the breaking up of As-MY aggregates in the decalin and hydrogen donation to the asphaltene become easier than the case only with decalin. Results of SANS experiment indicated that there exists the large aggregates (fractal network) in the case of Maya asphaltene with decalin up to 350 °C.[5] The fractal network may be related with the reaction properties of As-MY that have thermal cracking that have more coke yield than As-KF and As-IL, and that shows synergy effect by quinoline and decalin to suppress the coking.

Gas yields. In the case of As-MY and As-IL, the gas yields decrease much by the solvents. One possible hypothesis is as follows. Aggregations of As-MY and As-IL are stabler than that of As-KF.[6] This indicates the molecules in the aggregates of As-MY and As-IL are associated each other more tightly than that of As-KF. Consequently, the larger radical fragments produced by the thermal cracking in the aggregates of As-MY and As-IL tend to recombine more to make coke precursors than that of As-KF, and the smaller one of As-MY and As-IL tend to be released more as gas components than As-KF.

Conclusion

The reactivities of vacuum residues and asphaltenes are investigated with autoclave thermal cracking experiments at 440 °C. The results shows that the relaxation of asphaltene aggregates by agitation for one hour at 300 °C is effective to suppress the coking in the case with the solvents. The mixture solvent consisted with quinoline and decalin perform better than both the single solvent for As-MY, while sole decalin is better than the mixture solvent for As-KF and As-IL.

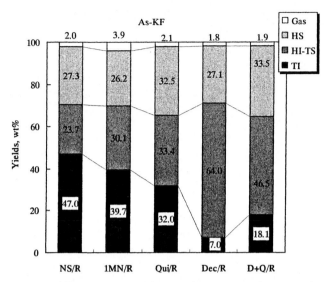

Figure 5. Yields of As-KF cracking without and with some solvents.

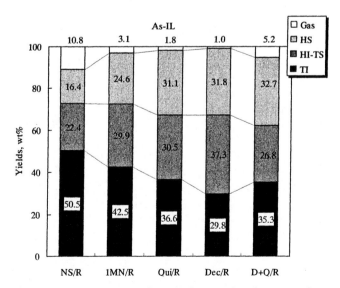

Figure 6. Yields of As-IL cracking without and with some solvents.

It is inferred from those results that the coke yields in heavy oil thermal cracking could be decreased by relaxation of asphaltene aggregates with an appropriate thermal and solvent treatment. The control of the solvent abilities of aspahltene dispersion and hydrogen donation is important to be suitable for each asphaltenes.

Acknowledgment: This work was supported by the Proposal-Based International Joint Research Program, New Energy and Industrial Technology Development Organization (NEDO).

References

1. Altgelt, K. H.; Boduszynski, M. M. *Composition and Analysis of Heavy Petroleum fractions;* Marcel Dekker: New York, 1994; Chapter 2.
2. Kidena, K.; Usui, K.; Murata, S.; Nomura, M.; Trisnaryanti, W. *J. Jpn. Petrol. Inst.* **2002,** *45,* 214-221.
3. Sato, S.; Takanohashi, T.; Tanaka, R. *Prepr. Pap.-Am. Chem. Soc., Div. Fuel Chem.* **2001,** *46 (2),* 353.
4. Takanohashi, T.; Yan, Z.; Sato, S.; Saito, I.; Tanaka, R. *Energy Fuels* **2003,** 17, 135-139.
5. Tanaka, R.; Hunt, J. E.; Winans, R. E.; Thiyagarajan, P.; Sato, S.; Takanohashi, T. *Energy Fuels* **2003,** *17,* 127-134.
6. Tanaka, R.; Seifert, S.; Hunt, J. E.; Winans, R. E.; Sato, S.; Takanohashi, T. In *Proceedings of 3rd International Conference on Petroleum Phase Behavior & Fouling held in conjunction with the AIChE 2002 Spring National Meeting;* New Orleans, U.S.A., March 10-14, 2002.
7. Mochida, I.; Moriguchi, Y.; Korai, Y.; Fujitsu, H.; Takeshita K. *Fuel* **1981,** *60,* 746-747.
8. Mochida, I.; Otani, K.; Korai Y. *Fuel* **1985,** *64,* 906-910.
9. Mochida, I.; Kishino, M.; Sakanishi, K.; Korai, Takahashi, R. *Energy Fuels* **1987,** *1,* 343-348.
10. Mochida, I.; Yufu, A.; Sakanishi, K.; Korai, Y. *Fuel* **1988,** *67,* 114-118.
11. Sakanishi, K.; Towata, A.; Mochida, I.; Sasaki, M. *Energy Fuels* **1988,** *2,* 802-807.
12. Mochida, I.; Takayama, A.; Sakata, R.; Sakanishi, K. *Energy Fuels* **1990,** *4,* 81-84.
13. Mochida, I.; Takayama, A.; Sakata, R.; Sakanishi, K. *Energy Fuels* **1990,** *4,* 585-588.

14. Constant, W. D.; Price, G. L.; McLaughlin, E. *Fuel,* **1986,** *65,* 8-16.
15. Chiba, K.; Tagaya, H.; Sato, S.; Kobayashi, T. *Proc. 1985 Int. Conf. Coal Sci.,* **1985,** 71-74.
16. Kamiya, Y.; Nagae, S. *Fuel* **1985,** *64,* 1242-1245.
17. Curtis, C. W.; Guin, J. A.; Tarrerr, A. R. *P0998A US DOE Rep.* **1989,** *DOE-PC-88801-T-3,* 20.
18. Clarke, J. W.; Rantell, T. D.; Snape, C. E. *Fuel* **1984,** *63,* 1476-1478.

Chapter 13

Thermal Processing Limits of Athabasca Bitumen during Visbreaking Using Solubility Parameters

Parviz M. Rahimi[1], Alem Teclemariam[1], Ed Taylor[1],
Theo deBruijn[1], and Irwin A. Wiehe[2]

[1]National Centre for Upgrading Technology (NCUT), 1 Oil Patch Drive,
Suite A202, Devon, Alberta T9G 1A8, Canada
[2]Soluble Solutions, 3 Louise Lane, Gladstone, NJ 07934

Conversion of Athabasca bitumen to marketable products requires a significant degree of upgrading. For instance, for pipeline transportation the bitumen's viscosity has to be substantially reduced either by addition of a diluent or thermal conversion. Petroleum feedstocks are stable in nature, but if the fine balance in their chemical composition changes because of physical or thermal treatment, there is a possibility of solid formation due to asphaltenes precipitation. If the pipeline viscosity requirement is met by the addition of diluent, the diluted bitumen has to be not only stable compatible with other crudes entering the pipeline. On the other hand, if the viscosity reduction is accomplished by thermal treatment then the presence of asphaltenes and the formation of high molecular weight components usually lead to coke and this has severe implications for refinery processing. Significant coke formation can occur in heat exchangers, furnaces and fractionators during petroleum processing. It is important, then, to be able to predict the coking onset of different process streams and advise operators to avoid process conditions (temperatures and space velocity) resulting in coke formation.

Introduction

The production of heavy oils and bitumen in Alberta is increasing and a significant portion will be transported to the US by pipeline. However, the transportation of these extra-heavy oils by pipeline is difficult because of their relatively high viscosity (>100,000 cP). The diluents used to lower viscosity for meeting the Canadian pipeline specification (350 cP) have significant paraffinic characteristics. The paraffinic character of the diluent can play a significant role in product instability by causing the precipitation of asphaltenes during transportation. Moreover, diluents are expensive and may not be readily available; pretreatment to improve viscosity and reduce diluent usage is an attractive option. Viscosity reduction of heavy oils can be accomplished by a number of processes, such as visbreaking and solvent deasphalting. Thermal processing of bitumen such as visbreaking for pipeline transportation produces liquid products that can be unstable and may cause solid and sediment formation. Compatibility and stability of fuels during transportation are important issues that have been given considerable attention in recent years.

For full conversion of bitumen and heavy oils into transportation fuels, in refineries it is necessary to subject these materials to varying degress of thermal treatment. In processes where relatively high temperatures are required, solid deposits may be formed as coke that significantly reduces the efficiency of the processing unit. For instance, in the delayed coking operation for processing Athabasca bitumen, the material is heated to temperatures in the range of 350-500°C (1). At these temperatures and in the absence of hydrogen, significant coke deposits can be formed on the walls of furnaces. These deposits currently can be removed by a pigging technique (1).

It is believed that petroleum is a colloidal system consisting of asphaltenes cores dispersed in solution by a polar fraction, namely resins. During thermal reaction this protective resin layer is destroyed and the asphaltenes become exposed. They are no longer soluble in the media, resulting in precipitation and, finally, coke is formed. The structure of asphaltenens themselves are also changed during thermal treatment for instance, side chain fragmentation and dehydrogenation reactions result in an increased aromaticity. There are limited data available regarding product stability in the reactor during thermal processing of bitumen. Viscosity reduction by solvent deasphalting may also result in products that are unstable and form sediment while being transported.

There are a number of methods and techniques that have been used to measure the stabilities and fouling tendencies of petroleum feedstocks and products:

1. spot test (ASTM-D-4740-95);
2. total sediment (ASTM-4870-96);
3. solubility parameters, optical microscope (2);
4. light scattering (PORLA) (3);
5. peptization value (P-value) (4);
6. colloidal instability index (CII) (5-6);
7. coking index (7).
8. Asphaltenes Stability Index (ASI) (8)

In the present work we adapted solubility parameters, through an optical microscopy method that was developed by Wiehe (2), to determine the coking onset of Athabasca bitumen. The main objective of this project was to extend the severity of the process beyond coke formation by using a batch autoclave, and then use a correlation between solubility parameters and severity to predict the coking onset. If the onset of asphaltenes precipitation or the coking onset of petroleum materials could be predicted during thermal processing, it would then be possible to avoid processing severities at which coke formation occurs. Data on the coking onset are also necessary for the design of different processing units.

In the present work the coking onset of Athabasca bitumen was determined using a batch autoclave. Using the solubility parameters of the total liquid products, the onset of coke formation was determined. It was shown that coke formation was negligible up to a pitch conversion (conversion of resid to distillate) of 34wt%. Properties of the total liquid products as a function of severity are discussed.

Experimental

Thermal reactions were performed using a 300 mL autoclave equipped with an insert (sleeve) for ease of removal and transfer of the products after the reaction. Feedstock used in this work was Athabasca bitumen obtained from Newalta, and originating from the "Dover SAGD project" (previously UTF) operated by Dover Canada. To perform thermal cracking, about 100 grams of the bitumen was warmed to 40°C and transferred to the autoclave sleeve. The sleeve was inserted into the 300 mL autoclave and the head tightened down using a torque wrench. The autoclave was then purged with nitrogen three times and pressure tested to 550 psi.

The excess pressure was reduced down to 100 psi. The autoclave was then insulated and its content was soaked to 150°C for one hour while stirring. The reactor was heated at 2.5-5°C/min to the final reaction temperature of 370-430°C. The severity index (SI) of the reaction was calculated according to the following equation:

$$SI = t*exp(-(Ea/R)*(1/T-1/700)) \qquad (1)$$

[where t = reaction time, seconds; Ea = activation energy, taken as 50.1 kcal/mole; R = gas constant, 0.001987 kcal/(mole °K); and T = reaction temperature, °K]

Upon achieving the pre-determined severity, the furnace was shut down and the insulation was removed. A cooling purge was directed to the head of the autoclave. The maximum internal temperature, the maximum pressure and the final severity index were recorded. Note that the severity indexes reported earlier as part of this work (9) were erroneously high and are subsequently corrected for this paper.

The autoclave was allowed to cool to room temperature and the gaseous products were collected in a gasbag (Calibrated Instruments Inc.) through a condenser. The condenser was cooled using dry ice to collect the light ends. The volume of the gas in the gasbag was determined by a gasometer with the barometric pressure and temperature recorded at that time. The contents of the gasbag were then analyzed for its components using a gas chromatograph (MTI Quad-series refinery gas analyzer).

The liquid products in the sleeve as well as the small amount of liquid collected in the reactor (as a result of overflow or condensation) were mixed. The light ends that were collected in the condenser were added to this mixture to produce total liquid products (TLP). The material balance (gas and TLP) and conversion of resid (+524°C) to distillate (-524°C), defined as pitch conversion, are shown in Table I. Determination of boiling point distribution of TLP was determined using high temperature simulated distillation. The properties of TLP, including solubility numbers and stability indexes are shown in Table II.

Table I. Product yields as function of severity

SI, second	Liquid wt%	Gas wt%	Pitch Conv * wt%
129	98.1	0.20	5.4
130	98.9	0.04	5.4
153	99.2	0.26	5.4
171	99.2	0.31	ND
192	99.0	0.31	7.3
192	99.7	0.3	ND
213	99.1	0.10	9.1
225	98.0	0.3	ND
275	98.1	0.43	14.5
311	98.1	0.47	16.3
307	99.2	0.5	ND
348	99.6	0.51	15.4
384	98.0	0.7	ND
406	98.3	0.52	18.2
483	99.2	0.87	24.5
672	97.4	1.37	32.7
717	98.4	1.29	34.5
776	97.4	1.53	34.5

* Conversion of +524°C to distillate determined by GC simulated distillation
ND = not determined

Table II. Properties of total liquid products

Severity Index, s	Heptane* Insol	Toluene Insol	MCRT	Aromaticity $^{13}C_{NMR}$	VISC.-DYN cP @ 25°C	IN	SBN	P-value
Athabasca	wt%	wt%	wt%					
Bitumen	11.3	0.03	13.7	0.33	174000	33.2	101.5	3.1
129	12.4	0.06	13.4	0.43	54800	37.3	111.0	3.0
130	12.4	0.07	13.0	0.41	57100	37.3	111.0	3.0
153	12.6	0.04	13.2	0.40	46300	38.2	101.5	2.7
171	ND	ND	ND	ND	ND	38.6	104.3	2.7
192	12.4	0.04	12.3	0.38	22000	40.7	106.6	2.6
192	ND	ND	ND	ND	ND	40.7	106.6	2.6
213	12.7	0.02	13.1	0.39	19700	40.7	106.6	2.6
225	ND	ND	ND	ND	ND	41.2	104.6	2.5
275	11.5	0.03	14.2	0.40	11566	41.4	108.5	2.6
311	12.4	0.11	14.0	0.42	9713	42.1	105.4	2.5
307	ND	ND	ND	ND	ND	45.3	104.1	2.3
348	12.5	0.00	13.2	0.45	5810	47.1	102.7	2.2
384	ND	ND	ND	ND	ND	49.4	103.7	2.1
406	11.6	0.04	13.9	0.38	3540	51.3	102.5	2.0
483	12.9	0.04	14.9	0.39	2860	56.8	103.4	1.8
672	13.6	0.04	15.1	0.40	984	69.4	101.3	1.5
717	14.0	0.10	15.1	0.42	1686	72.1	102.3	1.4
776	13.5	0.13	15.4	0.46	750	75.6	101.4	1.3
1250	**18.95**	**4.55**				**102.5**	**114.8**	**1.12**
1379	**16.42**	**4.58**				**89.4**	**109.1**	**1.22**
1410	**17.61**	**5.94**				**102.2**	**112.2**	**1.1**

* Including Toluene insolubles
ND = not determined

Results and discussion

Process yields

During thermal treatment of petroleum in refinery units, a solid deposit or coke is often formed. It has been shown that coking onset coincides with the insolubility of converted asphaltenes (7) due in part to the conversion of resins that protect the asphaltenes (10). There is also an induction period for coke formation, which depends on the feedstock and the severity of the process (11). In order to determine the coking onset of Athabasca bitumen, this material was subjected to different severities shown in Table I. Using high temperature simulated distillation, conversion of +524°C to distillate (pitch conversion) was calculated (Table 1). Up to a severity of 776 seconds, there was no coke (toluene insolubles) formed. At this severity, the pitch conversion, which is defined as [(+524°C in) − (+524° out)]/(+524°C in) was about 34wt%. These results are consistent with our previous data on the visbreaking of Athabasca bitumen in a continuous bench-scale unit where no coke was formed up to a pitch conversion of around 33wt% (12).

Product quality

The properties of the TLP are shown in Table II. Although there is some scatter in the data, in general, as the severity of the process increased the heptane asphaltenes (not corrected for the small amount of toluene insolubles) increased. Toluene insolubles (coke) did not increase with severity up to 776 seconds and examination of TLP under a microscope showed the presence of no solid particles or coke. Again, as expected and accepting the scatter in the data, both MCR and aromaticity increased with severity. A significant reduction in viscosity was observed as the severity increased.

Solubility parameters

The solubility numbers including S_{BN} (solubility blending number) and I_N (insolubility number) as determined by Wiehe's method (2) are shown in Table 2. These values were determined according to equations 1 and 2, respectively.

$$S_{BN} = I_N\left(1 + \frac{V_H}{5}\right) \qquad (2)$$

$$I_N = \frac{T_E}{\left(1 - \frac{V_H}{25d}\right)} \tag{3}$$

In these equations V_H is the maximum n-heptane that can be blended with 5 ml of oil without precipitation of asphaltenes and T_E is the minimum vol% toluene in the test liquid (toluene + n-heptane) required to keep asphaltenes in solution at 1 g oil and 5 ml of test liquid.

In a previous study (13) it was suggested that by using solubility numbers, it might be possible to predict the onset of coke formation for petroleum feedstocks during visbreaking experiments. The experiments were carried out in a continuous bench scale unit and in the presence of steam with severities below the threshold of coke formation. In the current experiments using a batch autoclave the severity of the experiments were extended beyond coke formation in order to confirm that the solubility parameters of the total liquids could be used to predict the onset of coke formation. In Figure 1, S_{BN} and I_N are plotted against the severity of the process. From the least square straight-line equations obtained for S_{BN} and I_N, it is possible to predict that at a severity index of 1147 seconds asphaltenes will become insoluble ($S_{BN} = I_N$) and start forming coke. Using this value in the SI equation (equation 1), the coke induction period at different temperatures can be determined. For instance, the coke induction period at 427°C for Athabasca full-range bitumen was calculated to be around 20 minutes. The predictability of coke formation was further investigated in a series of autoclave runs that were performed at a relatively high severity, as shown at the bottom of Table 2. Although more data needed for severities between 776 and 1147 seconds, as predicted, all experiments that were carried out with severities higher than 1147 seconds did produce coke. For example, thermal cracking of Athabasca bitumen at a severity of about 1250 seconds produced 4.5wt% coke, and the C_7 asphaltenes (not corrected for coke or TI) also increased significantly. It is expected that at the end of the coke induction period (i.e., at the severity where coke starts forming) the asphaltenes concentration will be reduced (7). For this reason the percentage of C_7 asphaltenes, after correcting for toluene insolubles, should have been lower than what is reported in Table 2. For the last three runs, at high severities beyond the coke induction period, it was necessary to filter out the solid particles (coke) in order to determine the solubility numbers (S_{BN} and I_N). These solubility numbers do not follow the linear behavior of those within the coke induction period because the least soluble asphaltenes precipitated to form coke.

Product stability

The stability of total liquid products (P-value) is plotted against the severity in Figure. 2. It can be seen that beyond the severity of 776 seconds the stability changed only slightly. The P-values, which are indicative of the stability of total liquid products, are calculated from the ratio of S_{BN}/ I_N and are also shown in Table 2. For the petroleum products to be stable, S_{BN} has to be greater than I_N or the ratio of S_{BN}/I_N, which is equivalent to P-value, has to be greater than 1.0. As the severity of the process increased the stability of products decreased because the converted asphaltenes became less soluble (higher I_N) and because the solvent quality became slightly poorer (lower S_{BN}). It should be mentioned however, that the stability of total liquid products at severities above 776 could only be determined after the removal of solid coke in these samples. It is also important to note that the insolubility numbers for the last three experiments (Table II) were significantly higher compared with those at lower severity. Although the P-values of these samples may indicate that they are stable, if they were to be blended with other crude, there could be compatibility problems with these liquids.

Product Compatibility

From solubility parameter measurements it was clear that as the severity increased the insolubility of the total liquid products increased. This may have some implication for blending these liquids with other crudes during pipeline transportation and may result in incompatibility and, finally, asphaltene precipitation. For any given mixture to be compatible, the criteria in equation 4 should be satisfied:

$$S_{BNMIX} > I_{NMAX} = \text{Maximum } I_N \text{ in Mixture} \tag{4}$$

According to this equation the solubility blend number of the mixture should be higher than the maximum insolubility in the mixture. The solubility blending number of the mixture of blends can be calculated using equation 5.

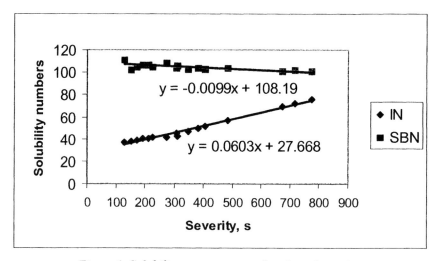

Figure 1. Solubility parameters as function of severity
(Reproduced with permission from *ACS Fuel Chemistry Preprints* **2003**, *48(1),* 103. Copyright 2003 by the authors.)

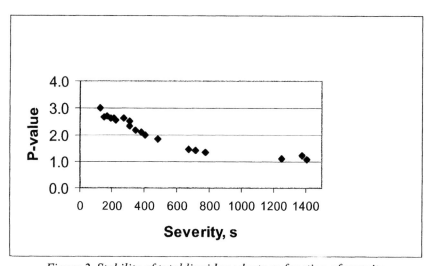

Figure 2. Stability of total liquid products as function of severity

$$S_{BN_{blend}} = \frac{V_1 S_{BN1} + V_2 S_{BN2} + \ldots V_n S_{BNn}}{V_1 + V_2 + \ldots V_n} \tag{5}$$

In order to determine whether the visbroken liquids from Athabasca bitumen are compatible with Alberta crudes, the solubility numbers of a number of Alberta crudes were determined as shown in Table III. The results indicate that all three crudes tested are stable and the Cold Lake blend has the highest stability. For comparison, in Table IV solubility numbers of the crudes are listed again with visbroken Athabasca bitumen. This visbroken liquid was selected because the severity of the reaction was on the threshold of coking and the amount of TI was relatively small. The compatibility of Alberta crude and visbroken liquids at a 50/50 ratio are shown in Table V. According to equation 5, the calculated solubility blending numbers for each blend should be higher than the insolubility number of 75.6 determined for visbroken liquids. The results shown in Table 5 indicate that, at the selected ratio, only Cold Lake crude is compatible with the selected visbroken liquids and the other two are either borderline or incompatible. It can be concluded that the selection of process conditions for visbreaking of bitumen and heavy oil (for the purposes of producing pipelineable liquids) should be such that not only the highest liquid yield are obtained but also that the liquid products are compatible with the blending crude. Even if compatible at the tested ratio, one should not blend the visbroken oil into a crude with a solubility blending number below the insolubility number of the visbroken oil, or risk precipitating asphaltenes because of blending in the wrong order (2).

Conclusions

Thermal reaction of Athabasca bitumen was carried out in a batch autoclave at SI ranging from approximately 130-1410 seconds. It was shown that up to a severity index of around 776 seconds there was no solid or coke formed. At this severity, pitch conversion was estimated to be about 34wt%. From the solubility numbers of the total liquid products versus the severity, it was possible to predict the onset of coke formation for Athabasca bitumen. For pipeline application, however, the selected severity of the process should produce liquid products that are compatible with crudes entering the pipeline.

Table III. Compatibility and stability of Alberta crude

Alberta crude	V_H (mL)	T_E (%)	I_N	S_{BN}	P-value
Sour blend	2.9	29.0	33.6	53.1	1.58
Sweet blend	4.0	21.0	26.0	46.8	1.80
Cold Lake	7.1	25.0	35.9	86.9	2.42

Table IV. Solubility numbers of Alberta crudes and visbroken liquids

Solubility numbers	Sour blend	Sweet blend	Cold Lake	Visbroken Athabasca
I_N	33.6	26.0	35.9	75.6
S_{BN}	53.1	46.8	86.9	101.4

Table V. Solubility numbers of Athabasca visbroken liquids with Alberta pipelineable crudes

Blend Ratio	Sour blend Visbroken (50/50)	Sweet blend Visbroken (50/50)	Cold Lake Visbroken (50/50)
S_{BN} of Blends	77.2	74.1	94.1

Acknowledgements

Partial funding for NCUT has been provided by the Canadian Program for Energy Research and Development (PERD), the Alberta Research Council (ARC) and the Alberta Energy Research Institute (AERI).

References

1. *"Mitigation of fouling in bitumen furnaces by pigging"*, Proceedings of 1st international conference on petroleum phase behavior and fouling, Parker, R. J., and McFarlane, R.A., AIChE Spring National Meeting, March 14-18, Houston, Texas, 412-418, 1999.

2. *"The oil compatibility model and crude oil incompatibility"*, Wiehe, I.A. and Kennedy, R.I. *Energy & Fuels*, 14, 56, 2000.

3. *"Experience in use of automatic heavy fuel oil stability analyzer"*, Proceedings of the 6th international conference of stability and handling of liquid fuels, Vilhunen J, Quignard A, Pilviö O, and Waldvogel, J. October 13-17, Vancouver, BC., 985-987, 1997.

4. *"Stability of visbroken products obtained from Athabasca bitumen for pipeline transportation"*, Rahimi, P.M., Parker, R.J. and Wiehe, I,A. Symposium on heavy oil resid compatibility and stability, 221st ACS National Meeting, San Diego, CA, USA, April 1-5, 2001.

5. *"Contribution a la Connaissance des proprietes des bitumen routiers"* Gaestel, C., Smadja, R. and Lamminan, K.A.. Rev Gen Routes Aerodromes, 85, 466, 1971

6. *"Petroleum stability and heteroatoms species effects on fouling heat exchangers by asphaltenes."* Asomaning, S.A. and Watkinson, A.P. *Heat Transfer Engng*, 12, 10-16, 2000.

7. *"A Phase-Separation Kinetic Model for Coke Formation"*, Wiehe, I.A. I&EC Research, 32, 2447-2454 (1993).

8. *"Crude oil blending effects on asphaltenes stability in refinery fouling"*, Samuel Asomaning, Petroleum Science and Technology, 21, 3-4, 569-579, 2003.

9. *"Determination of coking onset of petroleum feedstocks using solubility parameters"*, Symposium on heavy hydrocarbon resources-characterization, Rahimi, P.M., Alem, T., Taylor, E., DeBruijn, T. and Wiehe, I,A. upgrading and utilization, Division of Fuel Chemistry, 225th ACS National Meeting, New Orleans, LA, USA, March 23-27, 2003, 103-105

10. *"Predicting coke formation tendencies,"* Schabron, J.F., Pauli, A.T., Rovani Jr., J.F. and Miknis, F.P. *Fuel*, 80, 1435-1446, 2001.

11. *"A Series Reaction, Phase-Separation Kinetic Model for Coke Formation from the Thermolysis of Petroleum Resid"*, Wiehe, I.A. presented at the AIChE Spring Meeting, New Orleans (1996).

12. *"Visbreaking of heavy oils for pipeline transportation – quality and stability"*, Parker, R.J., Lefebvre, B. and Pipke, A. presented at the 2nd International conference on petroleum and gas phase behavior and fouling, August 27-31, 2000, Copenhagen, Denmark

13. *"Stability and compatibility of partially upgraded bitumen for pipeline transportation"*, Rahimi, P.M., Parker, R.J., Knoblauch, R. and Wiehe, I.A., Proceedings of the 7th International conference on stability and handling fuels, September 22-26, 2000, Graz, Austria, 177-192.

Chapter 14

Characterization of Soluble Mo Complex Catalyst during Hydrotreatment of Vacuum Residues

Kinya Sakanishi[1], Izumi Watanabe[2], and Isao Mochida[2]

[1]Institute for Energy Utilization, National Institute of Advanced Industrial Science and Technology (AIST), Tsukuba, Ibaraki 305–8569, Japan
[2]Institute for Material Chemistry and Engineering, Kyushu University, Kasuga, Fukuoka 816–8580, Japan

Soluble Mo complex catalysts prepared from Mo-DTC(dithiocarbamate) and Mo-DTP(dithio-phosphate) are characterized during the hydrotreatment of vacuum residues, in order to clarify how active species of Mo sulfides are formed and behaved under the hydrotreatment conditions. Far-Infra Red (IR) and Raman spectroscopy, and TEM observation showed that Mo-S species and MoS_2 lamellar structure are formed above 300 °C, and are higly dispersed to interact with asphaltene macromolecules for their effective hydrocracking.

Introduction

Highly dispersed Mo sulfide catalysts with fine particles have been reported very active for the hydrotreatment of heavy hydrocarbons such as vacuum residue(VR) and coal.[1-6] Several oil-soluble Mo complexes can be one of the most promising candidates for the application to slurry phase catalytic upgrading of vacuum residues, because they form very fine MoS_2 particles during their heating upto around 350 °C.[7] It is reported that the difference in the catalytic activities of Mo-DTC(dithiocarbamate) and Mo-DTP(dithio-phosphate) can be ascribed to the different extent of MoS_2 formation from the complexes under the hydrotreatment conditions. It is also noted that MoS_2 or solid Mo compounds derived from the two complexes are fine particles of very low crystallinity that can be highly dispersed in vacuum residue or asphaltene, because such solid particles are allowed to stay in asphaltene micelle which is dissolved or dispersed in toluene or maltene during the heating. Thermal decomposition of Mo-DTC stoichiometrically gave MoS_2, whereas Mo-DTP appears to provide MoS species in the presence of remaining phosphorous ligands. It seems that the remaining phosphorous ligands may retard the conversion of MoS to MoS_2 in VR, however, no clear evidence is not found by XRD, TGA, far-IR or TEM measurements.

In the present study, Raman spectroscopy is applied to the identification of MoS species formed from the soluble Mo complexes, since it can distinguish the layered graphite and amorphous-like carbon species. Because of the similarity in the layered structure of MoS_2 and graphite-like carbon, Raman spectroscopy is expected to give a clue to the difference in the structures of MoS species produced from the two Mo complexes.

Experimental

Materials. Mo-DTC and Mo-DTP used as MoS catalyst precursors in the present study are commercially available in the forms of crystal and solution, respectively. Their structures are illustrated in Fig.1. Commercially available MoS_2 powder is used as a reference compound. An Arabian heavy vacuum residue(AH-VR) is used for the hydrotreatment with Mo-complexes.

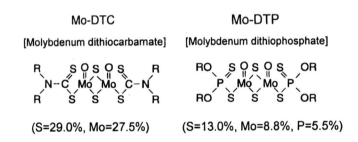

Mo-DTC [Molybdenum dithiocarbamate] (S=29.0%, Mo=27.5%)

Mo-DTP [Molybdenum dithiophosphate] (S=13.0%, Mo=8.8%, P=5.5%)

Fig.1 Structures of Mo-DTC and Mo-DTP

Heat-treatment. The mixture of Mo complex with VR was heated at 110 °C for 1 h for well dispersion, and then hydrotreated at 380 °C over the decomposition temperature of the complexes in a mini-autoclave under 10 MPa

H_2. The heated mixture was extracted with n-hexane to separate the maltene and asphaltene fractions.

Analyses. The thermal behavior of the Mo complexes was examined using TGA(Seiko, SSC/5200) by their weight changes. The complexes and their decomposed products in the residue were analyzed by FT-Far IR(Jasco-620), XRD(Rigaku Geigerflex), and Raman spectroscopy(Nippon Bunko NRS2000).

Results and Discussion

Characterization of soluble Mo complexes during heattreatment

According to TG/DTA profiles of Mo-DTC, three endothermic peaks were observed, and the first two of them were derived from the structural change in the ligands with a small weight change. The peaks around 339 and 478 cm^{-1} disappeared above 260 °C was ascribed to the decomposition of Mo-DTC, and above 300 °C, formation of MoS$_2$ with a large weight loss was observed as sharpened peak at 384 cm^{-1}. This was in good agreement with the formation of Mo-S bond observed by far IR spectroscopy in Fig.2. Such thermal decomposition behaviors of Mo-DTC were also described in a previous paper.[7]

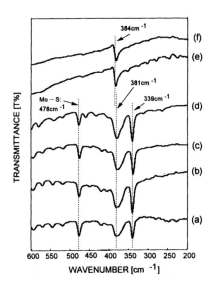

Fig.2 Far IR spectra of Mo-DTC before and after the heat-treatment
(a: non-treated, b: 200°C, c: 250°C, d: 260°C, e: 300°C, f: 400°C)

The decomposition of Mo-DTP was started from 200 °C based on the TG/DTA, reaching to 26 wt% at around 300°C, however, no clear peak derived from Mo-S species was observed in the far IR as shown in Fig.3.

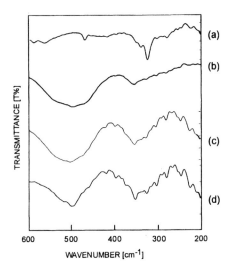

**Fig.3 Far IR spectra of Mo-DTP before and after the heat-treatment
(a:non-treated, b:350°C, c:400°C, d:500°C)**

Raman spectroscopy has been reported as one of the powerful tools for the direct and in-situ observation of the formations of Mo-S and Co-Mo-S species in the hydrodesulfurtization catalysts.[8,9] It is revealed that the formation of Mo-S bond in MoS_2 compounds can be detected by the two peaks around at 386 cm^{-1} (E^1_{2g} noramal mode) and 412 cm^{-1} (A_{1g} noramal mode) in Raman spectroscopy. Such peakes were observed in the heat-treatment of a commercial $CoMo/Al_2O_3$ catalyst under 10 % H_2S/H_2 atmosphere at 400 °C by in-situ Raman spectroscopy measurement.[9]

Fig.4 illustrates Raman spectra of Mo-DTP complex before and after the heat-treatment at variable temperatures under N_2 or 10 % 10 % H_2S/H_2 atmosphere. When Mo-DTP was heat-treated under N_2 atmosphere, no peak was observed at 350 and 500 °C, suggesting the thermal stability of Mo-DTP under the inert condition. On the other hand, When it was heat-treated under 10% H_2S in H_2, although no peak was observed at 270 °C, the formation of Mo-S species was clearly observed at 350 °C, showing the two peaks around at 386 cm^{-1} (E^1_{2g} noramal mode) and 412 cm^{-1} (A_{1g} noramal mode) in Raman spectroscopy of Fig.4.

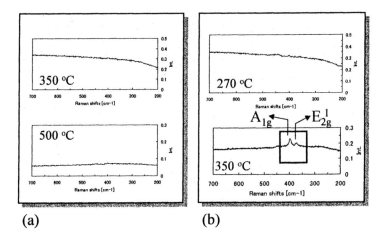

(a)   (b)

Fig.4 Raman spectra of Mo-DTP before and after the heat-treatment
(a: under N₂, 1h; b: 10%H₂S/H₂, 1h)

It is revealed that MoS_2 should be produced from Mo-DTP over its decomposition temperature under the sufficient sulfiding conditions. It is pointed out that Mo-DTC produced MoS_2 at around 300°C even under the N_2 atmosphere as illustrated in Fig.5.

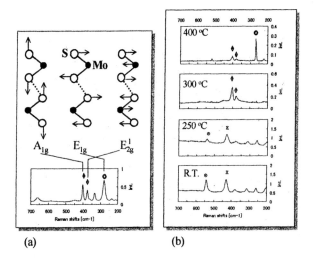

(a)   (b)

Fig.5 Raman spectra of Mo-DTC before and after the heat-treatment
(a: MoS₂; b: heat-treated Mo-DTC under N₂)

202

Catalytic activities of MoS species formed from Mo-DTC and Mo-DTP

The difference in the properties and crystallinity of MoS$_2$ between Mo-DTC and Mo-DTP may reflect in the Raman spectroscopy. Further details on the transformation of MoS$_2$ morphology and crystallinity during the hydrotreatment are now on investigation   Fig. 6 shows TEM micrograph of Mo-DTC heat-treated with asphaltene at   370 °C. A clear formation of lamella-like MoS$_2$ crystalline structure was observed on the asphaltene matrix.

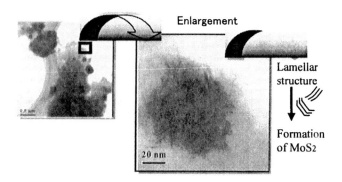

Fig.6 TEM of Mo-DTC treated at 370°C with asphaltene

Fig.7 illustrates TOF-MAS spectra of asphaltene after the hy   reatment of VR with or without Mo-DTC or Mo-DTP. Compared to the spectra without catalyst, both of the Mo complexes hydrocracked the asphaltene farction significantly, the cracking extent being a little higher with Mo-DTP.

According to the hydrotreatment results of VR with 50 wt% tetralin solvent at 380 °C for 2 h under 10 MPa of H$_2$ pressure, the yield of the asphaltene was lower with Mo-DTC than that with Mo-DTP, reflecting the higher hydrogenation activity of Mo-DTC, which was evaluated by the composition of tetralin-derived products after the hydrotreatment.

The yiled of long-chain paraffins after the hydrotreatment of VR was slightly higher with Mo-DTC catalyst than that with Mo-DTP, probably because the hydrogenation-assisted hydrocraking of alkylated aromatic hydrocarbons may be more favorable for Mo-DTC-derived MoS$_2$ catalyst.

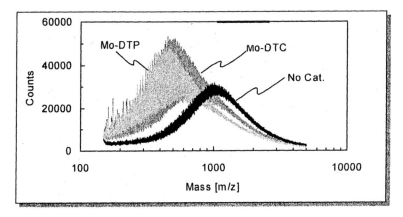

Fig.7 TOF-MAS spectra of asphaltene after the hydrotreatment with or without Mo complex catalysts

References

1. Strausz, O.P., Lown, E.M., Mojelsky, T.W., *Prepr.ACS Div. Petrol. Chem.*, 1995, 40(2), 741.
2. Del Bianco, A., Panarti, N., Marchionna, M., *Prepr.ACS Div. Petrol. Chem.*, 1995, 40(2), 743.
3. Fixari, B., Peureux, S., Elmouchnino, J., Le Perchec, P., *Energy Fuels*, 1994, 8, 588.
4. Del Bianco, A., Panariti, N., Di Carlo, S., Beltram, P.L., Carniti, P., *Energy Fuels*, 1994, 8, 593.
5. Panariti, N., Del Bianco, A., Del Piero, G., Marchionna, M., Carniti, P., *Appl.Catal. A: Gen.*, 2000, 204, 203.
6. Panariti, N., Del Bianco, A., Del Piero, G., Marchionna, M., Carniti, P., *Appl.Catal. A: Gen.*, 2000, 204, 215.
7. Watanabe, I., Otake, M., Yoshimoto, M., Sakanishi, K., Korai, , Y., Mochida, I., *Fuel*, 2002, 81, 1515.
8. Muller, A., *Polyhedron*, 1986, 5, 323.
9. Muller, A., Weber, T., *Appl. Catal.*, 1991, 77, 243.
10. Watanabe, I., Sakanishi, K., Mochida, I., Yoshimoto, M., *Prepr.ACS Div Fuel Chem.*, 2003, 48(1), 94.

Chapter 15

Novel Utilization of Mesoporous Molecular Sieves as Supports of Metal Catalysts for Hydrocracking of Asphaltene

Enkhsaruul Byambajav and Yasuo Ohtsuka[*]

Research Center for Sustainable Materials Engineering, Institute of Multidisciplinary Research for Advanced Materials, Tohoku University, Katahira, Aoba-ku, Sendai 980–8577, Japan

Formation of maltene from petroleum asphaltene at 573 K in pressurized H_2 proceeds more efficiently over 10 % Fe catalyst supported on mesoporous molecular sieves (SBA-15) with average pore diameters of 12 – 15 nm than on those with 4.6 – 7.5 nm. Loading of 10 % Ni and 3 % Ni/15 % Mo instead of 10 % Fe on the SBA-15 support with the diameter of 12 nm decreases asphaltene conversion but improves maltene yield. The 10 % Ni catalyst thus shows the highest selectivity in maltene formation among all the catalysts examined. Sulfur conversion increases with increasing asphaltene conversion, irrespective of the kind of catalyst, whereas sulfur conversion is larger with the Fe than with the Ni. The characterization of toluene insoluble fractions recovered after hydrocracking reveals that the Fe shows higher degree of dispersion on the support than the Ni, and that fine Fe-S species and surface Ni-S phases may play a crucial role in hydrocracking of asphaltene under the present conditions.

The importance of catalytic upgrading of heavy oils and petroleum residues to produce middle distillates has recently increased due to strict regulations on sulfur and particulate matter emissions and declining reserves of sweet crude oils. Asphaltene, the heaviest oil component defined usually as a fraction soluble in toluene but insoluble in n-heptane, is significantly included in heavy residues and frequently responsible for catalyst deactivation in the hydrotreating processes (1-4). According to earlier research, asphaltene molecules readily form micelles and aggregates with the sizes of more than 2 nm (5-8), and the mesopores (defined as the pores with the diameters of 2 – 50 nm) in catalyst supports can allow the micelles and aggregates to access to catalytically active species and consequently avoid catalyst deactivation (2, 9-13). Iron and Ni-Mo catalysts supported on USY (14) and HY (15) zeolites with newly formed mesoporous structures, respectively, have been developed for hydrocracking of atmospheric residues. All of the mesopores mentioned above are irregularly spaced and the sizes are broadly distributed.

On the other hand, mesoporous molecular sieves recently developed, such as MCM-41 (16), FSM-15 (17), and SBA-15 (18), have well-defined periodic mesopores, consequently provide very narrow pore size distributions, and possess large pore volumes of $1 - 2$ cm^3/g and high surface areas reaching 1000 m^2/g. These novel materials are expectable as the sites and spaces of adsorption, separation, synthesis and reaction for large molecules (19,20). It is therefore of interest to utilize such materials as catalyst supports for upgrading heavy residues. However, there has been only limited information about the use of MCM-41 and SBA-15 as supports of Ni-Mo and Co-Mo catalysts for hydrocraking of vacuum gas oils (21,22) and hydrodesulfurization of residual oils (23), and it has been suggested that the mesopores with the diameters of 2 – 5 nm are too small to crack asphaltene molecules (21-23). The SBA-15 support may be more suitable for this purpose, compared with the MCM-41, since the former can readily provide large pore diameters of more than 10 nm and possess high hydrothermal stability (18).

We have thus been working on the utilization of SBA-15 with different sizes of mesopores as supports of metal catalysts for cracking of asphaltene molecules under relatively mild conditions (24-27). The present paper first clarifies the relationship between average pore diameter of SBA-15 support and the performance of Fe catalyst supported on it, because there has been no systematic work about the effect of pore size of mesoporous molecular sieve on cracking behavior of asphaltene. Asphaltene conversion and maltene formation increase with increasing average pore diameter up to 12 nm but level off beyond this value, irrespective of the kind of gas atmosphere (25,26). The SBA-15 support with the diameter of 12 nm is thus selected, and some factors controlling the mild cracking of asphaltene in pressurized H_2 are then examined in detail. This paper finally describes not only the changes in crystalline forms of SBA-15-

supported catalysts before and after hydrocracking but also chemical states of catalyst components in reaction residues. Asphaltene itself separated from vacuum residues is used throughout this paper, because this may simplify interactions between asphaltene molecules and catalytically active species and may consequently be helpful to better understand the chemistry of asphaltene cracking.

Experimental

Synthesis of SBA-15 with Different Pore Diameters

Four kinds of SBA-15 were synthesized according to the method reported earlier *(18,28)*. The synthesis method has been reported in detail elsewhere *(28)* and is thus simply explained below. First, 1, 3, 5-trimethylbenzene (TMB) as a pore-swelling agent was added to an aqueous solution of a triblock copolymer ($EO_{20}PO_{70}EO_{20}$: EO, ethylene oxide; PO, propylene oxide) as a template and hydrochloric acid, weight ratio of $TMB/EO_{20}PO_{70}EO_{20}$ being in the range of 0.5 – 1.0. Then, the acidic mixture was heated at 308 K during stirring. After tetraethyl orthosilicate as a SiO_2 source was added to the mixture, the resulting gel was stirred at 308 K for 24 h and finally subjected to post-synthesis heat treatment at 370 – 380 K for 24 – 96 h.

As-synthesized SBA-15 was heated in a stream of air at 1 K/min up to 773 K and calcined for 6 h to remove the template. Four kinds of calcined SBA-16 with average pore diameters of 4.6 – 15 nm were prepared as catalyst supports, and the SBA-15 with the diameter of 12 nm was mainly used, unless otherwise described.

Addition of Catalyst Components to SBA-15 Supports

The addition of Fe, Ni or Mo component to the SBA-15 support was carried out by the conventional impregnation method with an ethanol solution of $Fe(NO_3)_3 \cdot 9H_2O$, $Ni(NO_3)_2 \cdot 6H_2O$ or $MoO_2(CH_3COCHCOCH_3)_2$ as a catalyst precursor, respectively. The last two compounds were co-impregnated with the support. In the impregnation process, one or two compounds were first dissolved into ethanol, and a predetermined amount of SBA-15 was then added to the ethanol solution. The resulting mixture was stirred at 313 K for 1 h, followed by removal of ethanol under vacuum at 313 K. The material recovered was calcined again in the same manner as above and then used as a catalyst. Each catalyst is denoted as Fe/SBA-15, Ni/SBA-15 or NiMo/SBA-15 throughout this paper. Loading of Fe or Ni metal in the former two catalysts was 10 wt %, and loading of Ni or Mo metal in the NiMo/SBA-15 was 3 wt % or 15 wt %, respectively, unless otherwise described.

A commercial Ni-Mo catalyst supported on SiO_2-Al_2O_3, denoted as NiMo/Si-Al, was also used as a reference, loading of Ni or Mo metal being 3 wt % or 15 wt %, respectively.

Feed Asphaltene and Preparation of a Mixture of Asphaltene and Catalyst

Feed asphaltene, which was defined as a toluene-soluble but n-heptane-insoluble fraction, was recovered from vacuum residue of crude oil from the Middle East by the precipitation method with 20-fold volume of n-heptane under ultrasonic irradiation. After a predetermined soaking time, the material precipitated was first separated by filtration and then dried under vacuum at 363 K. Yield of asphaltene recovered after dryness was 9.0 ± 0.5 wt % on the weight basis of the starting residue. The elemental analysis of asphaltene is: C, 84.0; H, 7.8; N, 0.9; S, 5.3; O (by difference), 2.0 wt%; Ni, 140; V, 400 ppm by weight.

In this work, for the purpose of holding asphaltene molecules within the mesopores of the SBA-15-supported catalyst, a toluene solution of asphaltene was first impregnated with the catalyst, and the resulting mixture was then held at room temperature under ultrasonic irradiation. After removal of toluene and subsequent dryness at 363 K, the mixture of asphaltene and the catalyst was finally used in all hydrocracking runs. The weight ratio of the two was 1.0, unless otherwise described.

Asphaltene Hydrocracking and Product Separation

All runs were carried out with a stirred stainless autoclave. Approximately 0.5 g of the mixture and about 1 g of tetralin as a solvent were first charged into the autoclave. After complete replacement with pressurized H_2 at the initial pressure of 5.0 MPa, the autoclave was heated at 5 K/min up to 573 K and then soaked for 1 h during stirring. Some runs without tetralin added were also performed for comparison.

Solid products after hydrocracking were separated by the Soxhlet method using n-heptane and toluene as solvents into three fractions, such as maltene (heptane-soluble), recovered asphaltene (heptane-insoluble but toluene-soluble), and coke plus catalyst (toluene-insoluble, denoted as TI throughout this paper). Most runs were repeated twice. Asphaltene conversion was obtained from the amounts of feed and recovered asphaltene. Maltene yield was calculated from the weight of maltene formed, and coke yield was estimated from the amount of TI fraction recovered on the basis of the assumption that the weight of every catalyst added is unchanged before and after hydrocracking. Asphaltene conversion and both yields are expressed in wt%.

Characterization

The N_2 adsorption measurements of SBA-15 supports and the supported catalysts were made by the conventional method at 77 K. The surface area was estimated by the BET method. The pore size distribution, pore volume, and average pore diameter were determined by the BJH method. The diameter was calculated by using the volume and total surface of mesopores *(18)*.

The powder X-ray diffraction (XRD) analysis was carried out by using Ni-filtered Cu Kα radiation to make clear the changes in crystalline forms and dispersion states of catalyst components before and after hydrocracking.

The reducibility of fresh Fe/SBA and Ni/SBA-15 catalysts after air calcination was examined by the temperature-programmed reduction (TPR) run, in which the sample was heated at 10 K/min up to 1273 K under flowing 67.5 vol% H_2/Ar, and the amount of H_2 consumed in this process was monitored on line with a thermal conductivity detector. The sulfur forms of TI fractions recovered after hydrocracking were analyzed by the temperature-programmed oxidation (TPO) method, in which the sample was heated at 3 K/min up to 1173 K in a stream of 10 vol% O_2/He, and the SO_2 evolved was on-line determined with a gas chromatograph equipped with a chemi-luminescence detector.

Results and Discussion

Pore Properties and Crystalline Forms of SBA-15 Supported Catalysts

Figure 1 shows typical examples of pore size distributions for SBA-15 support with average pore diameter of 12 nm and for the supported metal catalysts. Every support used in this work provided the asymmetric peak shape and narrow size distribution, which were characteristic of a mesoporous molecular sieve. The peak was broader at a larger average pore diameter. When metal components were added to the support, as shown in Figure 1, the peak shape and position did not change significantly, regardless of the kind of metal and the degree of its loading, but the peak height was always lower in the presence of the catalyst component. This means some decrease in pore volume by the incorporation of metal oxide particles inside the mesopores.

Pore properties and surface areas of all of SBA-15 supports and metal catalysts examined are summarized in Table I, where average pore diameter, pore volume, and surface area are denoted as D_p, V_p, and S_{BET}, respectively. The

D_p of the support was in the range of 4.6 – 15 nm and almost unchanged by catalyst addition, irrespective of the type of the metal added and the extent of its loading. The V_p of 10 % Fe/SBA-15 increased with increasing D_p and reached 2.4 cm^3/g at the largest D_p of 15 nm, though the S_{BET} was maximal at 7.6 nm. When the effect of metal loading on V_p was examined at almost constant D_p of 11 – 12 nm, the V_p decreased in the order of 4 % Fe ≤ 10 % Fe or 10 % Ni < 3 % Ni/15 % Mo < 30 % Fe, and it was thus the lowest at the highest metal loading of 30 %. The same tendency was observed with the S_{BET}. The lower V_p and smaller S_{BET} observed at a higher metal loading may mean that the corresponding amounts of catalyst particles are held inside the mesopores.

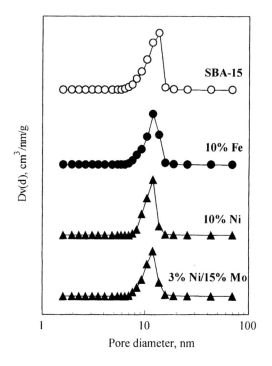

Figure 1. Typical examples of pore size distributions for SBA-15 support and metal catalysts with average pore diameters of 11 – 12 nm.

As shown Table I, the V_p and S_{BET} of the NiMo/SBA-15 were 1.9 cm^3/g and 530 m^2/g, respectively. These values were much larger than those (0.48 cm^3/g and 180 m^2/g) of the commercial NiMo/Si-Al. The pore size distribution of the NiMo/Si-Al revealed less developed mesoporous structures.

Table I. Pore Properties and Surface Areas of SBA-15 Supports and Fresh Metal Catalysts

SBA-15 support			Metal catalyst			
D_p [1] (nm)	V_p [2] (cm^3/g)	S_{BET} [3] (m^2/g)	Metal loading	D_p [1] (nm)	V_p [2] (cm^3/g)	S_{BET} [3] (m^2/g)
4.6	1.2	860	10 % Fe	4.5	0.97	700
7.5	2.2	900	10 % Fe	7.6	1.8	730
11.8	2.6	790	4 % Fe	11.6	2.3	800
			10 % Fe	11.8	2.3	680
			10 % Ni	11.7	2.1	620
			3 % Ni/15 % Mo	11.3	1.9	530
			30 % Fe	11.5	1.5	520
14.7	2.8	810	10 % Fe	14.9	2.4	650

[1] Average pore diameter. [2] Pore volume. [3] BET surface area.

Table II. XRD Results of SBA-15 Supported Catalysts and Changes in the Pore Volumes by Mixing with Asphaltene

Metal loading	D_p [1] (nm)	XRD species [2]	ΔV_p [3] $(cm^3/g\text{-}catalyst)$
10 % Fe	4.5	n.d. [4]	0.38
10 % Fe	7.6	n.d.	0.57
4 % Fe	11.6	n.d.	n.a. [5]
10 % Fe	11.8	n.d.	0.94
10 % Ni	11.7	NiO (s)	0.83
3 % Ni/15 % Mo	11.3	MoO_3 (w)	n.a.
30 % Fe	11.5	$\alpha\text{-}Fe_2O_3$ (w)	n.a.
10 % Fe	14.9	n.d.	1.0

[1] Average pore diameter. [2] Intensities designated as s, strong; w, weak.
[3] Decrease in pore volume by mixing with asphaltene.
[4] No species detectable. [5] Not analyzed.

Table II summarizes the XRD results of SBA-15-supported catalysts. Since all of Fe/SBA-15 catalysts were air-calcined at 773 K after impregnation using $Fe(NO_3)_3$ solution, the nitrate can be expected to be transformed to Fe_2O_3. However, no diffraction lines attributable to Fe species could be detected for the 4 and 10 % Fe catalysts with D_p of 4.5 – 15 nm. It is probable that Fe_2O_3

particles on all of these Fe catalysts are too small (less than 5 nm *(29)*) to be detected by XRD. As seen in Table II, on the other hand, very weak peaks of α-Fe_2O_3 were detectable with the 30 % Fe catalyst. In addition, strong XRD signals of NiO and small peaks of MoO_3 were observed with the Ni and NiMo catalysts, respectively. Although the diffraction lines of α-Fe_2O_3 and MoO_3 were too weak for the average crystalline sizes to be calculated by the Debye-Scherrer method, the average size of NiO was estimated to be 8.0 nm, which was lower than the D_p (12 nm) of the Ni catalyst. These observations show that all of these oxide particles are present inside the mesopores, irrespective of the kind of the metal, and that NiO is more readily crystallized than α-Fe_2O_3.

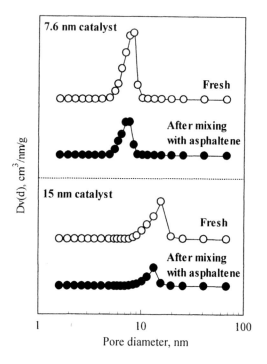

Figure 2. Changes in pore size distributions of Fe/SBA-15 catalysts before and after mixing with asphaltene.

Figure 2 shows typical changes in pore size distributions of two types of 10 % Fe catalysts before and after mixing with asphaltene in toluene. This mixing hardly changed the peak shape of the distribution, but it shifted the peak position slightly to a lower pore diameter and decreased the peak height considerably with all catalysts examined. Since it is reasonable to see that the asphaltene mixed is present inside the mesopores, the extent of the decrease in V_p by mixing, denoted as ΔV_P, may be used as a convenient index of the amount of the asphaltene held inside the pores. When ΔV_P calculated on the basis of asphaltene-free catalyst was compared among four kinds of 10 % Fe catalysts with different D_p, as shown in Table II, the ΔV_P was 0.38, 0.57, 0.94, and 1.0 cm^3/g at D_p of 4.5, 7.6, 12 and 15 nm, respectively. In other words, the ΔV_P increased with increasing D_p up to 12 nm, but it seemed to level off beyond this value. The ΔV_P for the 10 % Ni catalyst with D_p of 12 nm was roughly equal to that for the corresponding 10 % Fe catalyst. These observations suggest that the amount of asphaltene molecules held inside the mesopores after mixing may be higher at a larger pore diameter.

Performances of Several Catalysts in Asphaltene Hydrocracking

The determination of the amounts of recovered asphaltene, maltene and coke fractions reveals that material balances for all runs fall within the reasonable range of 95 – 107 wt %, and that the difference between asphaltene conversion and maltene yield can thus be regarded as coke yield in all runs.

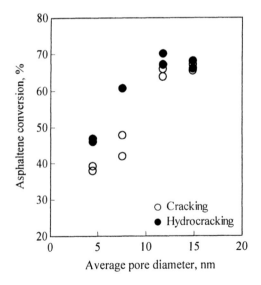

Figure 3. Dependency of asphaltene conversion on average pore diameter of 10 % Fe/SBA-15 catalyst.

The dependency of the conversion at 573 K on average pore diameter (D_p) of the 10 % Fe/SBA-15 catalyst is shown in Figure 3, where the data about the cracking in an atmospheric inert gas with a fixed bed quartz reactor *(25)* are also plotted for comparison. Asphaltene conversion during hydrocracking increased almost linearly with increasing D_p up to 12 nm and reached about 70 %, but it leveled off beyond 12 nm. This trend was also observed in the absence of H_2. The comparison of the results with and without pressurized H_2 shows that the conversion is larger in the hydrocracking using the Fe catalysts with the smaller D_p of < 10 nm, whereas it is almost independent of the atmosphere when the D_p is raised to $\geq$ 12 nm.

Figure 4 shows the effects of D_p on maltene yields in the presence and absence of H_2. The yield upon hydrocracking exhibited a quite similar dependency on the D_p as asphaltene conversion and provided the largest value of about 40 % at 12 – 15 nm, whereas the yield without H_2 did not change significantly with the D_p.

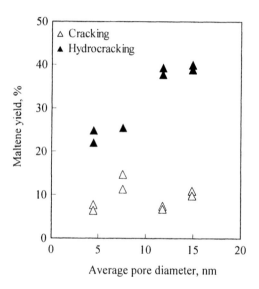

Figure 4. Maltene yield against average pore diameter of 10 % Fe/SBA-15 catalyst.

The relationship between asphaltene conversion and maltene yield in the cracking with Fe/SBA-15 catalysts is illustrated in Figure 5, where all of the data in Figures 3 and 4 are plotted. The yield was always larger upon hydrocracking when it was compared at the same level of the conversion. Since the difference between asphaltene conversion and maltene yield can be regarded as coke yield as mentioned above, it is evident that the coexistence of pressurized H_2 suppresses coke formation on the Fe catalyst remarkably, and that the extent of the suppression is much higher for the catalyst with a larger average pore diameter. The H_2 is thus essential for maltene formation from asphaltene molecules with the present Fe/SBA-15 catalyst.

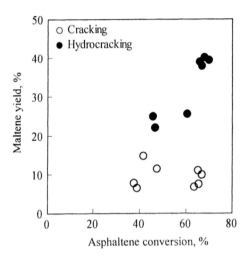

Figure 5. Relationship between asphaltene conversion and maltene yield with 10 % Fe/SBA-15 catalysts with average pore diameters of 4.5 – 15 nm.

The strong dependency of asphaltene conversion on the D_p of the 10 % Fe/SBA-15, observed in Figure 3, may be explained by the difference in the amount of asphaltene molecules held inside the catalyst mesopores in the process of mixing feed asphaltene with the Fe/SBA-15. The amount may be estimated by the decrease in the pore volume of the catalyst after mixing with the asphaltene. As mentioned earlier, the value (denoted as ΔV_P) calculated on the basis of asphaltene-free catalyst is provided in Table II. The ΔV_P increased monotonously from 0.38 cm^3/g at the D_p of 4.5 nm to 0.94 cm^3/g at 12 nm, but it seemed to level off beyond this value. This trend was quite similar to the relationship between the D_p and asphaltene conversion in Figure 3. It is thus probable that the higher conversion observed at a larger D_p originates from the presence of a larger amount of asphaltene molecules inside the mesopores of the Fe/SBA-15 catalyst.

As well known, the molecular size of asphaltene aggregates is dependent on the kind of organic solvent for them, the structure of asphaltene molecules, its concentration in the solvent, and temperature (5-7,30,31). The aggregate size of feed asphaltene in this work might be less than 12 nm under the present mixing conditions. Although almost all of the aggregates can be held inside the catalyst mesopores with the D_p of 12 nm or more, some of them may remain outside the pores with the D_p of 4.5 – 7.6 nm. Such a difference may be able to account for the dependency of asphaltene conversion on the D_p.

Since Figures 3 and 4 show that Fe/SBA-15 catalysts with larger D_p of 12 – 15 nm are suitable for the efficient conversion of the present asphaltene to maltene fraction, the 12 nm catalyst will be used mainly in further experiments.

Table III summarizes the effect of Fe loading in the SBA-15-supported catalyst with the D_p of 12 nm on asphaltene conversion and maltene formation during hydrocracking at catalyst/asphaltene ratio of 1.0. The conversion had the largest value of 75 wt % at 4 % Fe and decreased with increasing Fe loading. On the other hand, maltene yield was maximal at 10 % Fe. The most active 4 % Fe catalyst provided coke as the main product, whereas the 10 % Fe catalyst showed the highest selectivity (57 %) to maltene. The optimum loading for maltene formation thus existed under the present conditions. The relationship between Fe loading and asphaltene conversion will be discussed later.

Table III. Performances of Several Catalysts in Hydrocracking of Asphaltene

Support [1]	Metal loading	Catalyst to asphaltene [2]	Asphaltene conversion (wt%)	Maltene Yield (wt%)	Maltene Selectivity (%)
SBA-15	4 % Fe	1.0	75	22	29
SBA-15	10 % Fe	0.25	50	40	80
SBA-15	10 % Fe	0.50	54	42	78
SBA-15	10 % Fe	1.0	69	39	57
SBA-15	30 % Fe	1.0	58	25	43
SBA-15	10 % Ni	1.0	62	50	81
SBA-15	3 % Ni/15 % Mo	1.0	69	50	72
Si-Al	3 % Ni/15 % Mo	1.0	56	44	79

[1] SBA-15 with average pore diameter of 12 nm; Si-Al, commercial SiO_2-Al_2O_3.
[2] Weight ratio of catalyst/asphaltene.

The performances of Fe, Ni, and NiMo catalysts supported on the 12 nm SBA-15 in asphaltene hydrocracking are compared in Figure 6, where metal loading in the Fe or Ni catalyst is 10 wt %, and the result with the commercial NiMo/Si-Al, abbreviated as com-NiMo, is also shown for comparison. Material balances fell within very reasonable range of 101 - 105 %. Asphaltene conversion, maltene yield and selectivity observed for these catalysts are provided in Table III.

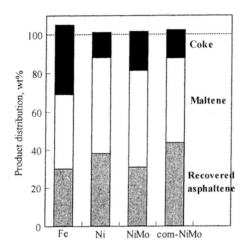

Figure 6. Product distribution in hydrocracking of asphaltene with several catalysts.

As seen in Figure 6 and Table III, asphaltene conversion increased in the order of com-NiMo < Ni < NiMo = Fe, whereas maltene yield was in the sequence of Fe < com-NiMo < Ni = NiMo, coke yield being maximal with the Fe catalyst. The Ni and com-NiMo catalysts thus provided the largest values of 79 – 81 % in maltene selectivity. Since mesopore volumes of the NiMo/SBA-15 and com-NiMo were estimated to be 1.9 and 0.48 cm^3/g, respectively, the much larger volume of the former catalyst may bring about higher asphaltene conversion and larger maltene yield than those over the commercial catalyst. The use of the 10 % Ni/SBA-15 in place of the 10 % Fe/SBA-15 lead to higher selectivity to maltene fraction. This difference may arise from the higher hydrogenation ability of Ni catalyst *(32)*. The observations in Figure 6 and Table III show that the 10 wt % Ni/SBA-15 is most suitable for maltene formation in the present hydrocracking among the four catalysts.

When the weight ratio of catalyst/asphaltene was reduced from the usual 1.0 to 0.50 and 0.25 while keeping the amount of asphaltene constant, as shown in Table III, asphaltene conversion during hydrocracking with the 10 % Fe/SBA-15 catalyst decreased from 69 % to 54 % and 50 %, respectively, whereas maltene yield remained almost unchanged (40 – 42 %). Thus, maltene selectivity at the smaller ratios of 0.25 – 0.50 was as high as 78 – 80 %, which was much higher than that (57 %) at the ratio of 1.0 and nearly equal to that observed with the 10 % Ni/SBA-15 catalyst. To decrease the amount of the Fe catalyst is therefore one of the key factors for selective formation of maltene from asphaltene with the Fe/SBA-15.

Tetralin was used as a solvent during hydrocracking, unless otherwise stated. To examine the solvent effect, the run without tetralin added was performed. Its absence did not affect asphaltene conversion significantly, whereas it decreased maltene yield and selectivity to some extent (26). The higher selectivity in the coexistence of tetralin may be ascribed to its hydrogen-donating ability, though the effect was small due probably to a low hydrocracking temperature of 573 K.

Catalyst States after Hydrocracking

As mentioned above, the 10 % Fe/SBA-15 showed larger catalytic activity for asphaltene conversion, and the 10 % Ni/SBA-15 provided the highest selectivity in maltene formation. Chemical states of these catalysts were thus examined mainly by the XRD and TPO measurements of TI fractions (coke plus catalyst) recovered after hydrocracking.

The XRD profiles are illustrated in Figure 7, where that for the mixture of feed asphaltene and the Fe or Ni catalyst before reaction is also shown for comparison. No diffraction lines of Fe species were detectable with the Fe catalyst, similarly as the case before hydrocracking. When Fe loading was increased to 30 %, as summarized in Table IV, very small peaks of Fe_3O_4 could be detected with the corresponding TI fraction, showing that α-Fe_2O_3 as the bulk species of the fresh 30 % Fe catalyst (Table II) is transformed to Fe_3O_4 but not completely reduced to metallic Fe even in pressurized H_2. With the Ni catalyst, on the other hand, weak but distinct diffraction lines attributable to metallic Ni appeared on the TI fraction, as seen in Figure 7. This means that partial reduction of the initial NiO to metallic Ni takes place during hydrocracking. Such a different reducibility of the Fe and Ni oxides was confirmed by the TPR runs of the fresh catalysts after air calcination (26). The TPR results also suggest that 4 – 10 % Fe catalysts in the hydrocracking process are present in more reduced forms than Fe_3O_4 observed with the 30 % Fe (26).

Although the XRD signals of any sulfided species could not be detected with the TI fractions, it can be expected that the surfaces of Fe and Ni catalysts are sulfided by H_2S evolved during hydrocracking. Figure 8 shows the typical profiles for SO_2 evolved in the TPO runs of TI fractions. The fraction with the 10 % Fe or 10 % Ni catalyst showed the main peak at 685 K or 720 K and the shoulder at 600 K or 560 K, respectively. These SO_2 peaks may be attributed to thiophenic and non-thiophenic forms in deposited coke on the catalysts *(33-36)*. This may show partial transformation of the former to the latter S-functionality in the hydrocracking process, because only a single peak of SO_2, identified as thiophenic forms, is observed in the TPO run of feed asphaltene alone *(25)*.

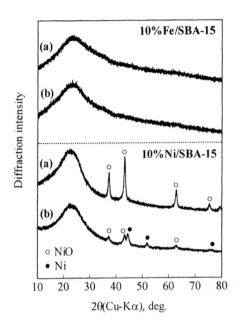

*Figure 7. Changes in XRD profiles before and after hydrocracking with 10 %
Fe/SBA-15 and 10 % Ni/SBA-15 catalysts: (a), a mixture of asphaltene and
catalyst before reaction; (b), a TI fraction recovered after reaction.*

Figure 8 also reveals the appearance of a small SO_2 peak around 875 K or 860 K with the Fe or Ni catalyst, respectively. The SO_2 may originate from the decomposition of Fe sulfates *(34,36)* and Ni sulfates *(37)*, since any SO_2 from the TI fraction without Fe or Ni metal added was not observed in this temperature region *(25)*. The 4 % Fe and 30 % Fe catalysts also showed the presence of the SO_2 evolved from the sulfates *(26)*.

The formation of Fe and Ni sulfates points out the presence of Fe-S and Ni-S phases in TI fractions, because it is likely that the sulfide species react with O_2 in the TPO run to form the corresponding sulfates, which are subsequently decomposed to evolve SO_2. Fe oxides, more reduced Fe species, NiO, or metallic Ni may be sulfided by reaction with H_2S evolved and/or directly with thiophenic sulfur forms in feed asphaltene during hydrocracking, according to the following equations:

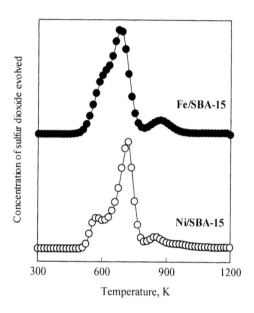

Figure 8. Profiles of SO_2 evolved during temperature programmed oxidation of TI fractions recovered after hydrocracking with 10 % Fe/SBA and 10 % Ni/SBA-15 catalysts.

$$Fe_3O_4 \text{ (and/or FeO, Fe)} + H_2S \rightarrow Fe\text{-}S \qquad (1)$$

$$NiO \text{ (and/or Ni)} + H_2S \rightarrow Ni\text{-}S \qquad (2)$$

To compare quantitatively the amounts of SO_2 evolved from metal sulfates in all TI fractions containing Fe/SBA-15 and Ni/SBA-15 catalysts, the peak area in Figure 8 was estimated by integrating the curve of the SO_2 profile between 800 – 1010 K. The results are summarized in Table IV, where the value is normalized to metal content in the TI fraction and expressed in arbitrary unit. Among the three Fe catalysts, the SO_2 evolved from the sulfates was larger at lower Fe loading, which means that the proportion of Fe-S species in the catalyst increases in the sequence of 30 % Fe < 10 % Fe < 4 % Fe. This order was the same as the dependency of asphaltene conversion on Fe loading at catalyst/asphaltene ratio of 1.0, as seen in Table III. Table IV also reveals that the amount of the SO_2 is lower with the 10 % Ni than with the 10 % Fe. This difference also corresponds to that in asphaltene conversion between the two (Table III). Since sulfided Fe and Ni catalysts promote hydrocracking of heavy residues (13,32), the results described above suggest that the sulfided species observed are catalytically active in the present reaction. As shown in Table IV, no diffraction lines of Fe-S and Ni-S phases were detectable in the TI fractions after hydrocracking, though the distinct XRD peaks of NiO and Ni could be detected compared with those of the Fe catalysts. It is thus probable that Fe-S and Ni-S species are present as very fine particles and at the outermost catalyst layers, respectively.

Table IV. XRD and TPO Results of Toluene Insoluble Fractions Recovered after Hydrocracking

Catalyst [1]	XRD species [2]	Amount of SO_2 evolved from metal sulfates [3]
4 % Fe	n.d. [4]	66
10 % Fe	n.d.	48
10 % Ni	NiO (w), Ni (w)	33
30 % Fe	Fe_3O_4 (vw)	19

[1] SBA-15 support with average pore diameter of 12 nm.
[2] Intensities designated as w, weak; vw, very weak.
[3] Normalized to metal content in TI fraction and expressed in arbitrary unit.
[4] No species detectable.

In hydrocracking of vacuum residue with presulfided Fe and Ni catalysts supported on activated carbon (32), the activity increased in the order of 1 % Ni < 10 % Fe < 1 % Fe, whereas yield of maltene or coke was maximal at 1 % Ni or 1 % Fe, respectively. These observations are quite similar to the present results provided in Table III, despite of the facts that feedstock, metal loading, and catalyst support are different between the two. It may be reasonable to see that supported metal catalyst is more finely dispersed at lower loading. Smaller particles of Fe sulfide species may thus show higher hydrocracking activity and promote conversion of asphaltene to coke to a larger extent. Maltene yield in the hydrocracking of the residue with the presulfided Fe catalyst was larger at loading of 10 % than at 1 % (32), whereas the yield in this work increased in the order of 1 % Fe ≤ 30 % Fe < 10 % Fe (Table III), showing that the hydrogenation ability of the present Fe is not necessarily larger at a higher loading. This might be ascribed to the lowest reducibility of the 30 % Fe *(26)* and the smallest proportion of the sulfide species in it (Table IV).

Fate of the Sulfur in Asphaltene during Hydrocracking

As the hydrocracking of asphaltene proceeds, conversion reactions of the sulfur in it also take place. It is of interest to examine the fate of the sulfur during hydrocracking with several catalysts. The sulfur remaining in recovered asphaltene and TI fraction (coke plus catalyst) was thus determined by the conventional elemental analyzer. Typical examples of sulfur distributions are illustrated in Figure 9, where the blank space is determined by difference and means the sulfur in maltene fraction (and/or H_2S). The sum of the sulfur in recovered asphaltene and TI fraction was almost 100 % with the 10 % and 30 % Fe/SBA-15 catalysts with average pore diameter of 12 nm. The same trend was observed with other Fe/SBA-15 catalysts used in the present work. In other words, almost all of the sulfur converted with every Fe catalyst went to each TI fraction. With the Ni/SBA-15 catalyst, on the other hand, some of the sulfur converted was included in maltene fraction (and/or as H_2S), as shown in Figure 9. This tendency was also observed with the NiMo/SBA-15 catalyst.

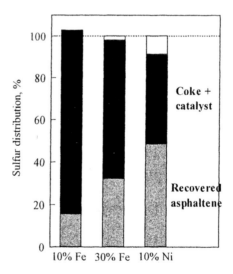

Figure 9. Typical distributions of the sulfur in asphaltene upon hydrocracking.

As shown in Figure 8, most of the S in TI fractions including Fe/SBA-15 and Ni/SBA-15 catalysts existed as thiophenic and non-thiophenic forms in coke fractions. The proportion of the latter S-functionality, which provided the shoulder peak of the SO_2 at 560 – 600 K in Figure 8, was larger with the Fe catalyst possibly because of higher asphaltene conversion.

The relationship between asphaltene conversion and sulfur conversion with SBA-15 supported catalysts with average pore diameters of 11 – 12 nm is provided in Figure 10, where the latter conversion is estimated by using sulfur contents in feed and recovered asphaltenes. As expectable, sulfur conversion increased with increasing asphaltene conversion, irrespective of the kind of catalyst. It should be noteworthy that all data for the Fe catalysts are plotted beyond the parity line between the two conversions, whereas the results with the Ni and NiMo catalysts are below the parity line. When catalytic hydrocracking of heavy oils was carried out with pressurized H_2 at 630 – 730 K, sulfur conversion always exceeded asphaltene conversion, and the degree seemed to be larger for vacuum residue with a higher content of asphaltene *(38)*. Figure 8 suggests that the Fe has stronger interactions with the sulfur in asphaltene than the Ni and NiMo, and that the removal of the sulfur from asphaltene occurs at the initial stage of the reaction with the Fe.

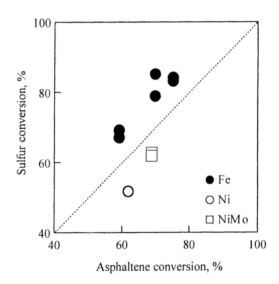

Figure 10. Relationship between asphaltene conversion and sulfur conversion during hydrocracking with SBA-15 supported catalysts with average pore diameters of 11 – 12 nm.

Conclusions

Hydrocracking of petroleum asphaltene mixed with Fe, Ni, and NiMo catalysts on SBA-15 supports with different average pore diameters of 4.6 – 15 nm has been studied with a stainless steel autoclave at 573 K under pressurized H_2. Asphaltene conversion and maltene yield on 10 % Fe/SBA-15 increase with increasing average pore diameter up to 12 nm but level off beyond this value. Loading of 10 % Ni and 3 % Ni/15 % Mo in place of 10 % Fe on the 12 nm SBA-15 support lowers asphaltene conversion, whereas it improves maltene yield, and consequently the Ni provides the highest selectivity in maltene formation. Conversion of the sulfur in asphaltene is higher with the Fe than with the Ni and NiMo, suggesting weaker interactions with the sulfur for the latter catalysts. The X-ray diffraction and temperature-programmed oxidation measurements of toluene-insoluble fractions after hydrocracking with the Fe and Ni catalysts show that the Fe is more highly dispersed, and that fine Fe-S species and surface Ni-S phases may be catalytically active.

Acknowledgement

This work was carried out partly as a research project of the Japan Petroleum Institute commissioned by the Petroleum Energy Center with a subsidy from the Ministry of Economy, Trade and Industry of Japan.

References

1. Yen, T.F.; Chilingarian, G.V. *Asphaltenes and Asphalts;* Yen, T.F.; Chilingarian, G.V., Eds.; Elsevier: Amsterdam, 1994; Vol. 1, p. 1.
2. Furimsky, E.; Massoth, F.E. *Catal. Today* **1999**, *52*, 381.
3. Kabe, T.; Ishihara, A.; Qian, W. *Hydrodesulfurization and Hydrodenitrogenation;* Kodansha & Wiley: Tokyo, 1999; p.325.
4. Sheu, E.Y. *Energy Fuels* **2002**, *16*, 74.
5. Yen, T.F. *Chemistry of Asphaltenes;* Burger, J.W., Ed.; Advances in Chemistry Series, 195, American Chemical Society: Washington, DC, 1981; p. 39.
6. Yen, T.F. *Encyclopedia of Polymer Science and Engineering;* Grayson, M.; Krochwitz, J.I., Eds.; Wiley: New York, 1988; p. 1.
7. Espinat, D.; Rosenberg, E.; Scarsella, M.; Barre, L.; Fenistein, D.; Broseta, D. *Structures and Dynamics of Asphaltenes,* Mullins, O.C.; Sheu, E.Y., Eds.; Plenum: New York, 1998; p. 145.

8. Artok, L.; Su, Y.; Hirose, Y.; Hosokawa, M.; Murata, S.; Nomura, M. *Energy Fuels* **1999**, *13*, 287.
9. Gawrys, K.L.; Spiecker, P.M.; Kilpatrick, P.K. *Prep. Am. Chem. Soc. Div. Petrol. Chem.* **2002**, *47*, 332.
10. Inoue, S.; Takatsuka, T.; Wada, Y.; Nakata, S.; Ono, T. *Catal. Today* **1998**, *43*, 225.
11. Sato, K.; Nishimura, Y.; Shimada, H. *Catal.Lett.* **1999**, *60*, 83.
12 Sakanishi, K.; Manabe, T.; Watanabe, I.; Mochida, I. *J. Jpn. Petrol. Inst.* **2000**, *43*, 10.
13. Terai, S.; Fukuyama, H.; Uehara, K.; Fujimoto, K. *J. Jpn. Petrol. Inst.* **2000**, *43*, 17.
14. Iino, A.; Iwamoto, R.; Nakamura, I. *Proc., First Tokyo Conference on Advanced Catalytic Science and Technology*, Tokyo, 1990; p. 351.
15. Sato, K.; Nishimura, Y.; Honna, K.; Matsubayashi, N.; Shimada, H. *J. Catal.* **2001**, *200*, 288.
16. Kresge, C.T.; Leonowicz, M.E.; Roth, W.J.; Vartuli, J.C.; Beck, J.S. *Nature* **1992**, *359*, 710.
17. Yanagisawa, T.; Shimizu, T.; Kuroda, K.; Kato, C. *Bull. Chem. Soc. Jpn.* **1990**, *63*, 988.
18. Zhao, E.; Feng, J.; Huo, Q.; Melosh, N.; Fredrickson, G.H.; Chmelka, B.F.; Stucky, G.D. *Science* **1998**, *279*, 548.
19. *Zeolites and Mesoporous Materials at the Dawn of the 21st Century;* Galarneau, A.; Di Renzo, F.; Fajula, F.; Vedrine, J., Eds.; Elsevier: Amsterdam, 2001.
20. *Nanoporous Materials III;* Sayari, A.; Jaroniec, M., Eds.; Elsevier: Amsterdam, 2002.
21. Corma, A.; Martinez, A.; Martinez-Soria, V.; Monton, J.B. *J. Catal.* **1995**, *153*, 25.
22. Ahmed, S. *Prep. Am. Chem. Soc. Div. Fuel Chem.* **2001**, *46*, 591.
23. Reddy, K.M.; Wei, B.; Song, C. *Catal. Today* **1998**, *43*, 272.
24. Byambajav, E.; Ohtsuka, Y. *Proc., 52nd Canadian Chemical Engineering Confererence*, Vancouver, 2002; p. 122.
25. Byambajav, E.; Ohtsuka, Y. *Fuel* **2003**, *82*, 1571.
26. Byambajav, E.; Ohtsuka, Y. *Appl. Catal. A: General* **2003**, *252*, 193.
27. Byambajav, E.; Tanaka, R.; Ohtsuka, Y. *Prep. Am. Chem. Soc. Div. Fuel Chem.* **2003**, *48*, 106.
28. Wang, Y.; Noguchi, M.; Takahashi, Y.; Ohtsuka, Y. *Catal. Today* **2001**, *68*, 4.
29. Radovic, L.R.; Walker, Jr., P.L.; Jenkins, R.G. *J. Catal.* **1983**, *82*, 382.
30. Roux, J.N.; Broseta, D.; Dem B. *Langmuir* **2001**, *17*, 5085.
31. Tanaka, R.; Winans, R.W.; Hunt, J.E.; Thiyagarajan, P.S.; Sato, S.; Takanohashi, T. *Prep. Am. Chem. Soc. Div. Fuel Chem.* **2001**, *46*, 359.

32. Nakamura, I.; Fujimoto, K. *Catal. Today* **1996,** *29,* 245.

33. Furimsky, E.; Y. Yoshimura, Y. *Ind. Eng. Chem. Res.* **1987,** *26,* 657.

34. LaCount, R.B.; Anderson, R.R.; Friedman, S.; Blaustein, B.D. *Fuel* **1987,** *66,* 909.

35. van Doorn, J.; Barbolina, H.A.A.; Moulijn, J.A. *Ind. Eng. Chem. Res.* **1992,** *31,* 101.

36. LaCount, R.B.; Kern, D.G.; King, W.P.; LaCount, Jr. R.B.; Miltz, Jr. D.J.; Stewart, A.L.; Trulli, T.K.; Walker, D.K.; Wicker, R.K. *Fuel* **1993,** *72,* 1203.

37. Rossini, F.D.; Wagman, D.D.; Evans, W.H.; Levine, S.; Jaffe, I. *Selected Values of Chemical Thermodynamic Properties;* National Bureau of Standards (U.S.): 1952; p.500.

38. Asaoka, S.; Nakata, S.; Shiroto, Y.; Takeuchi, C. *Ind. Eng. Chem. Process. Des. Develop.* **1983,** *22,* 242.

Indexes

Author Index

Subject Index